Practical Statistics for the Biological Sciences

Simple Pathways to Statistical Analyses

STEPHEN ASHCROFT AND CHRIS PEREIRA

palgrave
macmillan

First published 2003 by
PALGRAVE MACMILLAN
Houndmills, Basingstoke, Hampshire RG21 6XS and
175 Fifth Avenue, New York, N.Y. 10010
Companies and representatives throughout the world

PALGRAVE MACMILLAN is the global academic imprint of the Palgrave Macmillan division of St. Martin's Press, LLC and of Palgrave Macmillan Ltd. Macmillan® is a registered trademark in the United States, United Kingdom and other countries. Palgrave is a registered trademark in the European Union and other countries.

ISBN 0–333–96044–0

This book is printed on paper suitable for recycling and made from fully managed and sustained forest sources.

A catalogue record for this book is available from the British Library.

10   9   8   7   6   5   4   3   2   1
12   11   10   09   08   07   06   05   04   03

Printed and bound in Great Britain by

Creative Print & Design (Ebbw Vale), Wales

Contents

List of Figures

List of Tables

List of Boxes

List of Tutorials

Step-by-step tutorials

PractiStat tutorials

Preface

Biological scientists at all levels – undergraduate, postgraduate, post-doctoal – will sooner or later need to submit their data to statistical analysis. Many will have received little or no training in statistical methods. Others may have attended statistics courses but have had little hands-on experience in using the methods at the bench. This text is intended to be a handy guide to choosing and applying the appropriate statistical tests. It is compact enough to be used in the laboratory or in the field while providing sufficient detail to give the user some knowledge of the theoretical basis for the methods covered. A unique feature of the book is that it is accompanied by a CD-ROM with both Macintosh and Windows versions of a new programme, PractiStat, which can be used first to work through the examples given in the text, and second, as a tool for analysis of the user's own data. PractiStat provides a simple, intuitive interface, and permits the application of the common statistical tests and procedures used by bioscientists. Terms that are in **bold** type are defined in more detail in the Glossary (p. 157).

There are of course already a number of excellent statistical texts for the biosciences. However the in-depth treatment given by the best of these is more than the average researcher needs in order to subject data to analysis, nor do these texts typically provide a programme for carrying out the analyses. Such programmes do of course exist and many are indeed excellent. However they are often complex, with a rather steep learning curve, and again they may be more elaborate than the average bioscientist requires. Moreover they are also rather expensive for a limited undergraduate or postgraduate budget. We have therefore combined an accessible, concise, but nevertheless rigorous guidebook with an easy to use yet powerful statistical analysis programme that we hope will prove both useful and affordable for students and more senior workers in the biosciences.

Our thanks are due to those colleagues who kindly read the draft text, tested PractiStat, and made helpful criticisms and suggestions. These are David Harris, William James, David Raubenheimer, University of Oxford, UK, and José Vallo, University of Cadiz, Spain.

Stephen Ashcroft

Chris Pereira

Basic concepts of statistics

An introduction to the science behind 'significantly different'

Chapter 1 introduces the concept of measurement and the problems caused by variability in observed or collected data. We examine variation due either to experimental or measurement error or to intrinsic variation in the property/characteristic we are measuring, and show how statistics can describe and measure such variability, and determine its probability. We then examine the variables that can be found in a population – quantitative, qualitative, dependent and independent. Next we show how a statistical distribution can represent the outcomes of a given event in a bell-shaped normal curve after our population has been sampled. Finally, we show how to obtain a test statistic from our experimental data using the PractiStat software to conduct a statistical test, and how to handle any Type I or Type II errors in our tests.

1.1 Why use statistics?

The basis of science is *measurement* and all scientists seek to arrive at justified conclusions based on measured data. An inconvenient characteristic of all such measurements is that they contain variability that often prevents us from arriving at an absolute and unequivocal conclusion. Instead, we employ statistical analysis to arrive at a conclusion that, while not absolute, can be demonstrated mathematically to be *unlikely to be wrong*.

1.2 Variability in our data

In the life sciences, we often collect our data in the laboratory, in the clinic, in the greenhouse or out in the field. The variation we observe in our collected data can come from two sources:

1 Variation due to experimental or measurement error, and
2 Intrinsic variation in the property or characteristic that we are measuring.

Variation due to experimental error frequently occurs in our laboratories. For example, if you aliquot out $10 \mu l$ samples using a pipette, the actual volumes delivered will vary slightly from aliquot to aliquot because your operation of the pipette will not be perfectly repeatable from one sample to the next.

Variation is also an intrinsic property of biological characteristics. For example, if you measure the height of 10 women of the same age these values will not be identical because different women of the same age have different heights.

1.3 How does statistics handle variability?

Statistics handles variability in two ways. First it provides precise ways to *describe and measure the extent* of variability in our measured data. Secondly it provides us with methods for using those measures of variability to *determine a probability* of the correctness of any conclusions we draw from our data. Chapter 2 describes the ways to measure variability and subsequent chapters present various methods for us to arrive at justified conclusions about our data.

1.4 Observations, samples and populations

When we assemble a set of observations or measurements of a particular characteristic, we refer to this set of data as a **sample**. Although these data may be of interest in their own right, usually we are only interested in them for the information they can provide about the larger **population** consisting of *all possible* measurements of the characteristic in question.

In general it is impractical to obtain the complete set of all possible measurements on a population. Statistics therefore provides a means to assess how reliably we can extend the observations of our sample, to predict the results we would obtain if we were able to make observations on the *entire* population. The process of drawing conclusions about populations based on observations of a sample from the population is called **statistical inference**.

1.5 Variables

We use the term **variable** to describe a characteristic of a person or thing that we can measure, control or manipulate. Variables can be quantitative or qualitative. **Quantitative variables** record the *amount* of something; **qualitative variables** describe the *category* to which the data can be assigned and are therefore sometimes referred to as **categorical variables**.

1.5.1 Quantitative variables: continuous or discrete

Quantitative variables allow us to record an amount for each observation and to compare the magnitude of differences between them. They can be either *continuous* or *discrete*.

Continuous variables

These can take on any value in our infinite spectrum of real numbers. For example the concentration of cholesterol in a plasma sample may be 2.01, or 2.15, or $3.64\,g.l^{-1}$.

Discrete variables

These can take on only specific available values. For example the number of bacterial colonies on a Petri dish can only be a positive integer value; there can be 24 colonies, but never 24.5 colonies or –24 colonies.

1.5.2 Qualitative variables: nominal or ordinal

Qualitative variables allow us to have categories to which our observations may be assigned. These categories can be either *ordinal* or *nominal*.

Ordinal variables

These are assigned to groups or categories that can be placed in **rank order** on a scale of magnitude. An observation that is ordinal has no measured amount and is assigned to the category that best suits it. For example, after receiving a local anaesthetic, a patient may describe the pain of the injection as 'None', 'Mild' or 'Strong'. Although the categories in this example have no numerical value, it is clear that they can be ranked in increasing magnitude of pain from 'None' to 'Strong'.

Nominal variables

These are qualitative variables whose categories are completely arbitrary and cannot be meaningfully ranked. Examples of nominal variables include race, gender and blood type. It would make no sense to rank A-positive blood as 'bigger' than B-negative blood since the category is only a *description* of the blood and not an indication of magnitude.

1.5.3   Dependent and independent variables

The term **independent variable** is used to refer to a variable that is manipulated by the researcher, in contrast to the **dependent variable** whose value is affected by the manipulation – i.e. the independent variable influences the dependent variable but not vice versa. For example, if we measure the effect on blood pressure of different concentrations of a drug, then blood pressure is the dependent variable and drug concentration the independent. We can also use the terms 'dependent' and 'independent' to describe variables whose relationship can only plausibly be in one direction. For example an epidemiologist may obtain data relating the incidence of asthma to the level of particulates in the atmosphere. Clearly if there is a relation, it can only be that the level of atmospheric particulates influences the incidence of asthma; the converse could not be true. Hence the incidence of asthma would be the dependent variable and the particulates level the independent variable.

◀ 1.6 Statistical distributions

There is some degree of variability in every population and when we take a relatively small sample of observations from a larger population we can only *estimate* the characteristics of that larger population. So how do we arrive at justified conclusions based on our observations if there is uncertainty in our estimates? The answer is that we use a **statistical distribution** to determine the probability of occurrence of our particular sets of observations. To illustrate, consider the following case of the loaded dice.

Suppose we were watching someone throw a pair of dice and two sixes were thrown. Assuming we were familiar with the possible outcomes of throwing two dice, we would know that this combination was one of the less frequent outcomes, but we wouldn't necessarily accuse the thrower of being a cheat. Now suppose the thrower picked up the two dice, added a third and threw three sixes. We might have raised an eyebrow at this outcome because we might guess that throwing three sixes with three dice is an even less frequent occurrence than throwing two sixes with two dice. If the thrower then added a fourth dice and immediately threw four sixes, we might have been justified in concluding that the dice were loaded. This is because, unconsciously, we would have been comparing the observed outcomes to a distribution of *all possible outcomes* and coming to our justified conclusion that the probability of the dice *not* being loaded was unlikely.

A **statistical distribution** is a representation, either mathematical or tabulated, of all possible outcomes of a given event.

In the case of throwing two dice and summing their values, all the possible outcomes are shown in Table 1.1.

Table 1.1 Possible outcomes of throwing 2 dice

Die 1	Die 2	Sum	Die 1	Die 2	Sum	Die 1	Die 2	Sum
1	1	2	3	1	4	5	1	6
1	2	3	3	2	5	5	2	7
1	3	4	3	3	6	5	3	8
1	4	5	3	4	7	5	4	9
1	5	6	3	5	8	5	5	10
1	6	7	3	6	9	5	6	11
2	1	3	4	1	5	6	1	7
2	2	4	4	2	6	6	2	8
2	3	5	4	3	7	6	3	9
2	4	6	4	4	8	6	4	10
2	5	7	4	5	9	6	5	11
2	6	8	4	6	10	6	6	12

The statistical distribution is:

Sum of dice	2	3	4	5	6	7	8	9	10	11	12
No. of ways	1	2	3	4	5	6	5	4	3	2	1

or, graphically, as shown in Figure 1.1.

There are 36 possible outcomes, only one of which gives a total of 12. Hence the chance of throwing 12 is 1 in 36, or 0.0278.

In the case of throwing three dice and summing their values, the statistical distribution is shown in Figure 1.2.

The illustration shows that there are 216 possible outcomes when throwing three dice and that there is only one way a sum of 18 can be

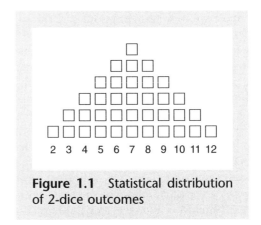

Figure 1.1 Statistical distribution of 2-dice outcomes

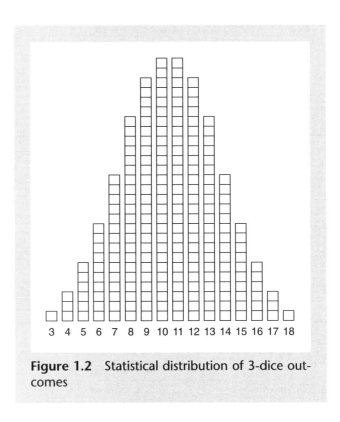

Figure 1.2 Statistical distribution of 3-dice outcomes

thrown (three sixes). Therefore the probability of throwing three sixes is 1 in 216, or 0.0046.

As we add more dice to be thrown, we may find that our distributions of possible outcomes become so large that we have to employ a computer to generate them. For a very large number of dice the shape of the graph-

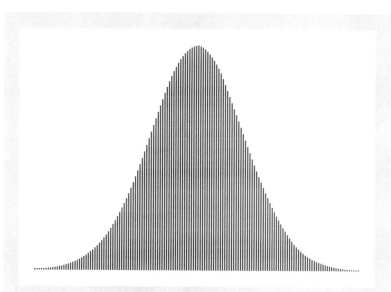

Figure 1.3 Statistical distribution of outcomes for a large number of dice

ical representation of the distribution might look like the bell-shaped curve in Figure 1.3.

1.7 The normal curve

In order to determine how meaningful an observed apparent relationship is in our samples, we would ideally have a function that tells us exactly how likely it is to obtain a relation of a given magnitude (or larger) from a sample of a given size by chance – i.e. what is the probability of observing a particular relationship in the sample if there is in fact no such relation between those variables in the population. The postulate that there is no relation in the population is called the **null hypothesis**. Such a function would give us the significance of the relationship, and would tell us the probability of error, P, involved in rejecting the null hypothesis. Although we usually don't know the precise form of function appropriate for our data, it is found empirically that many data can be fitted to functions related to a general type of curve called the **normal curve** (Figure 1.4).

The normal curve is a smooth, symmetrical bell-shaped curve defined by the following equation:

$$y = \frac{1}{\sigma\sqrt{(2\pi)}} e^{-\frac{1}{2}\left(\frac{x-\mu}{\sigma}\right)^2}$$

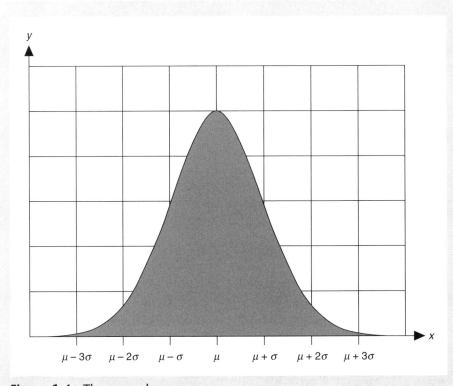

Figure 1.4 The normal curve

The exact shape of the normal distribution is defined by the two parameters μ and σ. The **population mean**, μ, is the midpoint of the distribution and the **population standard deviation**, σ, is a measure of the spread of the values about the midpoint. The concepts of mean and standard deviation are fundamental to statistical analysis and are discussed in detail in Chapter 2. If we define the total area under the curve as 100%, then the area under the curve from $x = \mu - \sigma$ to $x = \mu + \sigma$ is 68.26% of the total area. Similarly the area defined by $x = \mu \pm 2\sigma$ is 95.44% of the total; and the area defined by $x = \mu \pm 3\sigma$ is 99.74% of the total. To put it another way, 95% of the values can be calculated to lie within $\mu \pm 1.96\sigma$ and 99% within $\mu \pm 2.58\sigma$.

The significance of this curve for statistics is that the mathematician Gauss discovered that when large samples of measurements of a parameter are made, a frequency distribution of the resulting values – i.e. a plot of the frequency of occurrence of particular values – approximates closely to the normal curve.

Any observation, x, on the **abscissa** (the x-axis) can be expressed as a multiple of the population standard deviation. Such a **normalised value** is known as a **z-score**. Its utility is that we can then relate it to the general properties of all normal curves:

$$z\text{-}score(x) = \frac{x - \mu}{\sigma}$$

A value that lies 1 standard deviation greater than the mean is normalised to a value of +1. Since 95% of the values in a normal distribution lie within $\mu \pm 1.96\sigma$, the probability that a z-score greater than 1.96 could be obtained by chance is 0.05. Similarly, the probability of a z-score greater than 2.58 being due to chance is 0.01.

1.8 Random sampling

In making a statistical inference, the crucial question is how representative of the population is the actual sample that we choose. This is least likely to be a problem when the data represent measurements of an experimental variable in the laboratory under carefully controlled conditions using precise and specific assays – e.g. measurements of the rate of formation of $^{14}CO_2$ from [$U\text{-}C^{14}$] glucose by liver cells in vitro. However, where measurements are being made on groups of human subjects then the question of possible *bias* in the selection procedure has to be carefully assessed before it can be concluded that the sample is indeed representative of the larger population. For example, a sample of blood cholesterol levels obtained from 100 patients in Norway may not be representative of the blood cholesterol levels of our global population since there may be factors such as diet or climate which can affect cholesterol levels and are specific to Norway.

1.9 Probability and significance

To draw conclusions from experimental data we need first to set arbitrary critical thresholds of probability (**P-values**). The occurrence of an event whose estimated probability is less than a critical threshold is regarded as a **statistically significant** outcome. The usual **significance levels** chosen are $P < 0.05$ – significant; $P < 0.01$ – highly significant; $P < 0.001$ – very highly significant. As discussed above, we can use the properties of the normal curve to estimate probabilities. The procedure for deciding if an outcome is significant is called a **statistical test**.

1.10 Statistical tests

The objective of a statistical test is to obtain from the experimental data a single number called a **test statistic** whose probability distribution is known; the z-score is one such statistic. We then determine if the value of the test statistic exceeds some **critical value** associated with a particular probability threshold. For example, suppose we carry out a large series of measurements of the weight of a certain species of insect and find that the mean weight (μ) is 5.21 g with a standard deviation of 0.26 g. What is the likelihood that a randomly selected insect of weight 5.81 g could belong to the same population? The null hypothesis, H_0, is that this observation does come from the same population. The z-score is calculated as:

$$\frac{x - \mu}{\sigma} = \frac{5.81 - 5.21}{0.26} = 2.30$$

Since this value is greater than 1.96, the critical value of z for P = 0.05, we reject H_0 and conclude that the randomly selected insect is probably *not* from the same population.

When we set the threshold for rejection of H_0 at P = 0.05 we are effectively saying that we accept that there is a 5% chance that we shall reject H_0 when in fact it is true. Such an error is referred to as a **Type I error**, usually expressed as a probability and symbolised as α; in this case $\alpha = 0.05$. We can reduce the likelihood of committing a Type I error by setting more stringent conditions for rejection of H_0, e.g. P = 0.01. However the more we reduce the possibility of a Type I error the more likely we are to accept a null hypothesis that is in fact false: this is referred to as a **Type II error**, usually symbolised as β. The **power** of a test, $1 - \beta$, is the probability of the test reaching the correct conclusion – i.e. rejecting the null hypothesis when it is incorrect. The chance of making a Type II error can be lowered by increasing the number of observations in the sample. For some experiments, however, there may be severe financial or other constraints on the number of observations that can be included in the study. Hence it becomes important to calculate the minimum number of observations required to achieve the desired power for a given value of α. Such **power calculations**, the details of which vary depending on the test used, are complex and best performed with a specialised computer program such as G • Power.[1]

In this book we describe the most common statistical tests used by researchers in the biosciences.

1.11 Which test to use

It is important to choose the appropriate statistical test for your data. This is not always straightforward and indeed sometimes more than one test can

---

[1]G • Power is available as freeware for both Macintosh and PC (MS-DOS or Windows) from http://www.psycho.uni-duesseldorf.de/aap/projects/gpower/index.html.

be used. To guide you, we show in Boxes 1.1–1.3 a set of flow diagrams designed to lead you to the correct choice. If you are unfamiliar with the statistical terms used in Boxes 1.1–1.3 then we suggest you read Chapters 2 and 3 before attempting to use the flow diagrams.

Box 1.1: Which test to use?

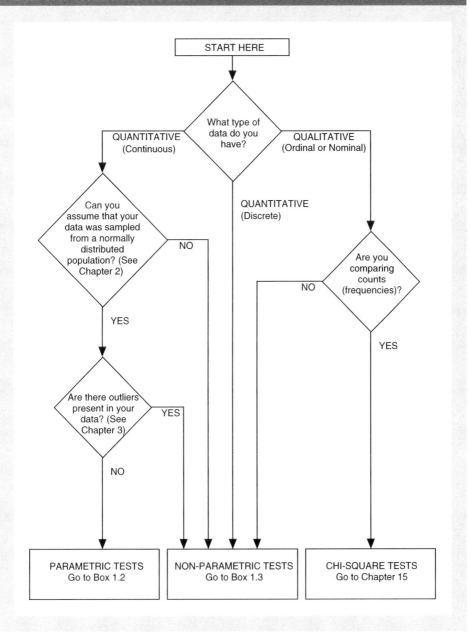

Box 1.2: Parametric tests

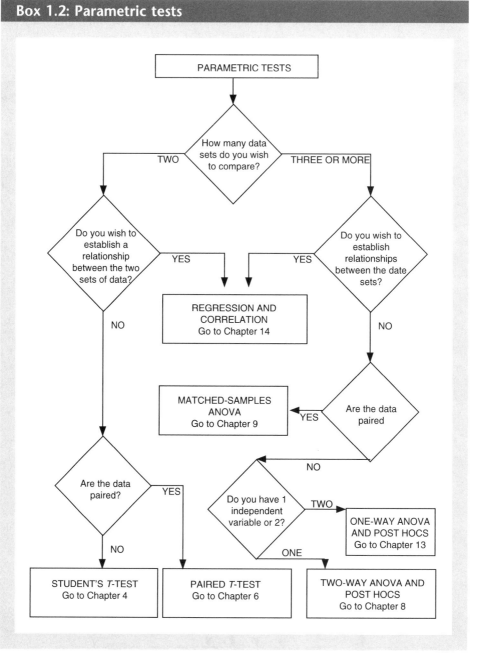

Box 1.3: Non-parametric tests

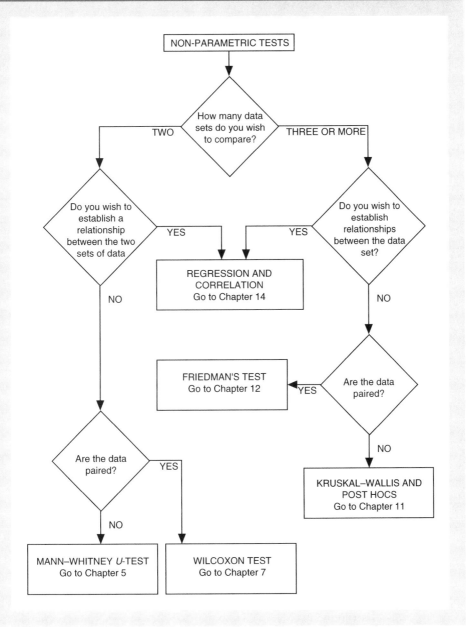

PractiStat tutorial 1.1: using PractiStat to perform statistical analyses

Throughout this book, worked examples in the form of Step-by-step tutorials of the various statistical tests are provided to enable you to carry out the tests from first principles. However a major feature of this book is the provision of a powerful yet simple computer programme, *PractiStat*, to help you analyse your data. Using PractiStat to carry out your statistical tests can be done in three easy steps:

1 Enter the data into PractiStat as individual samples

2 Select the statistical test you wish to perform

3 Select the samples you wish to test.

Step 1   Entering Data in PractiStat

PractiStat uses a data entry system that reflects the manner in which many data in the biological sciences are collected and recorded – i.e. 1 sample at a time. There are several ways to enter data in PractiStat: (a) type new data into an available sample space, (b) open existing data from a TEXT file (sometimes called an ASCII TEXT file); (c) if you feel more comfortable doing so, you may also enter your data into a familiar spreadsheet programme or word processor and simply 'Copy and Paste' to PractiStat.

(a) To *type data into PractiStat*, follow the three steps outlined below:

- When PractiStat is first opened, the interface defaults to data entry mode and presents an available sample space ironically titled 'Untitled 1'. If PractiStat has already been open and an available sample space is not presented, simply press the NEW button from the toolbar or select *New Data* from the *File* menu.

- An empty entry field (or edit field) will be active and ready to accept typed data. The data entry cursor (the flashing vertical bar) should be in the entry field and the text to the left of the entry field should read 'Observation #0001'. Type in the value of the first observation then hit the RETURN or ENTER key. Enter data for observations '0002', '0003', etc. in the same manner. An entry that begins with a non-numeric character will be shown in italics and excluded from all analyses.

- Once all of the sample's observations are entered, press the SAVE button or select *Save* from the *File* menu. Every sample is saved as an individual TEXT file containing that sample's observations. The name that is given to the saved file will replace 'Untitled 1' as the sample's name.

(b) To open existing data from a TEXT file, simply press the OPEN button on the toolbar or select *Open data file . . .* from the *File* menu. Use the presented navigation window to locate and open the file. The file's data will be shown in the observations list below the entry field and the file's name will shown in the samples list to the right of the observations list. We will be using this method to open the data files included on the PractiStat CD for the PractiStat tutorials of later chapters:

- To create a TEXT file using an alternate programme such as a word processor or spreadsheet, enter your data into a new document in columnar form. In spreadsheets, ensure that the data are entered in a single column and in word processors ensure that each value is on a separate line by following each entry with a RETURN keystroke. You can also enter lines of descriptive text which could help you identify or label your data. Non-numeric data will be ignored by the analysis routines in PractiStat. Use the programme's Save or Export command to save the completed file as a TEXT file.

(c) Instead of importing data from a TEXT file created as described above you can also copy a column of data from the spreadsheet or word processing programme (select *Copy* from the programme's *Edit* menu) and then paste it directly into an available sample space in PractiStat in data entry mode (select *Paste* from PractiStat's *Edit* menu).

Step 2   Selecting the statistical test

Once your data is entered in PractiStat, you simply choose the statistical test you wish to perform by pressing the appropriate button on the toolbar or selecting the appropriate menu item from the *Statistics* menu. If there is uncertainty as to which test is most appropriate, you can first consult the 'Which test to use? flow chart in Box 1.1.

When the desired statistical test is selected, the PractiStat interface will show an incomplete layout of the results of that analysis and will await your selection of the samples you wish to test.

Step 3   Selecting the samples you wish to test

To add samples to the analysis, simply click on the sample names from the samples list to the right of the PractiStat window. To select more than one sample, hold down the Control (CTRL) key on your keyboard (Mac users use the Command key) while clicking from the samples list. For each sample that is added, PractiStat will update the analysis and results layout to include the latest addition. To remove samples, hold down the Control (or Command) key and click on the sample to be removed.

A small icon that looks like a notepad is to the immediate left of each sample name in the samples list and indicates the status of the sample. A grey notepad indicates that the sample has no data, a closed-face (brown) notepad indicates that the sample has data and an open-faced notepad (blue horizontal lines) indicates that the sample has data and is included in the current analysis.

When testing for significance a statistically significant result will be indi-cated by 'P < 0.05' or by 'P = 0.05') in the P section of the results window and a non-significant result by 'NS (P > 0.05)'. The results of all statistical tests can be copied to the clipboard by selecting *Copy* from the *Edit* menu. and subsequently pasted into a document in another program by selecting *Paste* from that programme's *Edit* menu.

Each statistical test has its own requirements with respect to the amount of observations and numbers of samples needed to perform a successful analysis. A few tests even require you to flag a particular sample to distinguish it from the other samples in the analysis. When performing a statistical analysis in PractiStat for the first time, it is best to consult the chapter that describes the test and pay special attention to the instructions in the 'PractiStat Tutorial' section.

Editing of values in a selected data set is initiated by clicking the EDIT button or selecting *Edit Data* from the *Statistics* menu. Although PractiStat does not allow formatting of data in Edit mode, the number of decimal places shown in the analysis results window can be changed by selection of the formatting menu items in the *Special* menu.

1.12 Scope of PractiStat

PractiStat was specially created as a ready and easy visual tool for the analysis of all the statistical tests presented in this book (2-way analysis of variance excepted). The data from the tutorials are in TEXT file format in the 'Samples' folder on the PractiStat CD-ROM for your convenience. In addition to the tutorials in this book, PractiStat may be used as an instructional tool to analyse many field and laboratory data. It has enough data capacity to handle 25 samples with 1,000 observations per sample.

2

Descriptive statistics

Measures of central tendency and variability

Chapter 2 explains how we can use the concept of central tendency to represent the mode, the median or the mean values in our statistical sample. We then show how the sample mean can be used to indicate the variability of values in our data, and how to determine a confidence interval – a given level of certainty about the location of the population mean. We then examine common measures of variability – the range and interquartile range – and explain the concept of degrees of freedom. We next show how the coefficient of variation can be used to express the standard deviation in our data as a percentage of the mean. We finally examine the importance of the shape of a variable's distribution – how far it exhibits skewness or kurtosis relative to the bell-shaped curve of the normal distribution.

2.1 Measures of central tendency

Ask a person how long it takes them to get to work and we can expect an answer like, 'It takes me 35 minutes *on average*'. Without any formal exposure to statistics, most people already understand the concept of **central tendency** or typical score. Given five travel times for Monday to Friday, most people will use one of three methods to arrive at an answer to our question:

1 Of the five given times, they will choose the most frequently occurring, or **mode**, or

2 the middle or **median** magnitude of the five times, or

3 they will calculate the average or **mean** of the five times.

This idea that there is a particular value which most nearly represents the 'true' value is fundamental to statistics. The different ways of stating this central tendency are as follows.

2.1.1 The mode

The **mode** is the *most frequently occurring* measure in our sample. If, for example, our Monday to Friday travel times were 25, 30, 35, 35 and 40 minutes, we can say that our mode is 35 minutes since it occurred twice and all other times each occurred only once. However, finding the mode of a sample can be sometimes challenging if we are measuring a **continuous variable**. If we had measured our travel times to one decimal place of precision, we might have recorded times of 24.9, 30.2, 35.3, 34.8 and 39.6 minutes. In this case, the mode is not obvious and we would have to employ the use of a histogram to determine a measure of the mode (see Chapter 3).

2.1.2 The median

The **median** of a sample is the value of the *middle observation* when the observations are arranged in order from smallest to largest. If there is an odd number of observations, then the middle observation (sorted by magnitude) is our median. If there is an even number of observations, then our median is the average of the two middle observations. For example, consider the 8 numbers 2, 4, 6, 8, 10, 12, 14 and 16. Since the number of observations is even, the median is the average of the 4th and 5th observations sorted by magnitude; *median* = (8 + 10)/2 = 9.

2.1.3   The mean

The **mean** is the most commonly used measure of central tendency in statistics. It is the *arithmetic average* of our measures. If there are n observations in a sample and x_i is the value of the ith observation then the sample mean $\bar{x}$, pronounced 'x-bar', is given by the equation:

$$\bar{x} = \frac{\sum\limits_{i=1}^{n} x_i}{n}$$

Σx_i is the sum of all the observations in the sample

2.2   Measures of variability

The sample mean is of most use when we can give some indication of the variability of the values around it. The **confidence intervals** for the sample mean give us a range of values around the mean where we can state with a given level of certainty that the 'true' (population) mean is located (see the subsection *Confidence Interval for the Mean* below).

2.2.1   The range

The simplest way to express the deviation of the sample values from the mean is as a **range**. Thus for values of serum cholesterol of 2.40, 3.30, 2.50, 3.20, 1.89, 3.56, 4.05 g.l^{-1} the range would be the difference between the smallest and largest value – i.e. 4.05 – 1.89 = 2.16 g.l^{-1}. Although easy to calculate, the range is sensitive to extreme values and rarely used in statistical analysis.

2.2.2   The interquartile range

A more **robust** (i.e. less sensitive to extreme values) measure of variability is the **interquartile range**. This is the range of the middle 50% of observations when ranked by magnitude. To calculate this we first rank the observations by magnitude and subtract the first quartile value (Q1) from the third quartile value (Q3). A **quartile** is a group of ranked observations that represents one-quarter of a sample and the Q1 value is that number that the first quartile falls below. Similarly, the Q3 value is that number that the first three quartiles fall below. The interquartile range (IQR) is the difference between Q3 and Q1. Like the median, which incidentally is the Q2 or second quartile value, the Q1 and Q3 values may not be an actual observation in your sample, but may lie somewhere between two observations:

$$\text{Position of } Q1 = \frac{1}{4}n + \frac{1}{2}$$

$$\text{Position of } Q3 = \frac{3}{4}n + \frac{1}{2}$$

Interquartile range (IQR) $= Q3 - Q1$

where n is the number of observations

2.2.3 Percentiles

The **percentiles** of a sample of values divide the values into a hundred parts just as the quartiles divide the values into four parts. For example, the 10th percentile of a variable is a value such that 10% of the values of the variable fall below that value. Note that an alternative definition of the interquartile range of a variable is the value of the 75th percentile minus the value of the 25th percentile. Thus it indicates the width of the range about the median that includes 50% of the values:

$$\text{Position of } yth \text{ percentile} = \frac{y}{100}n + \frac{1}{2}$$

where n is the number of observations

Step-by-step tutorial 2.1: the interquartile range

Let us take the cholesterol data given above:

2.40, 3.30, 2.50, 3.20, 1.89, 3.56, 4.05 g·l^{-1}

Step 1 First rank the data in ascending order:

1.89, 2.40, 2.50, 3.20, 3.30, 3.56, 4.05 g·l^{-1}

Step 2 The minimum value is 1.89, the maximum value is 4.05 and therefore:

The **range** = maximum − minimum = 4.05 − 1.89 = 2.16

Step 3 For an odd number of observations the **median** is the middle value when ranked in order and therefore:

The **median** value is the 4th value, 3.20

Step 4 The **mean** value is (1.89 + 2.40 + 2.50 + 3.20 + 3.30 + 3.56 + 4.05)/7 = 2.99

Step 5 The first quartile position is:

$$\frac{1}{4}n + \frac{1}{2} = \frac{1}{4} \bullet 7 + \frac{1}{2} = 2.25$$

Another way of expressing this is: *the Q1 value lies 0.25 or one quarter of the way between the second and third observation.* Therefore:

$$\textbf{Q1} = 2.40 + 0.25(2.50 - 2.40)$$
$$= 2.40 + 0.25 \times 0.10$$
$$= 2.40 + 0.025$$
$$= 2.425\,g \cdot l^{-1}$$

Step 6 Similarly, the third quartile position is:

$$\frac{3}{4}n + \frac{1}{2} = \frac{3}{4} \bullet 7 + \frac{1}{2} = 5.75$$

$$\textbf{Q3} = 3.30 + 0.75(3.56 - 3.30)$$
$$= 3.30 + 0.75 \times 0.26$$
$$= 3.30 + 0.195$$
$$= 3.495\,g \cdot l^{-1}$$

Step 7 The **interquartile range** = Q3 − Q1 = 3.495 − 2.425 = 1.07 g·l⁻¹

2.2.4 Variance

The **variance**, σ^2, of a population of values (**population variance**) is computed as:

$$\sigma^2 = \frac{\sum (x_i - \mu)^2}{N}$$

Where μ is the population mean
  N is the population size

Since in general we cannot measure the entire population, the **unbiased estimate** of the population variance, s^2, is computed as:

$$s^2 = \frac{\sum (x_i - \bar{x})^2}{n - 1}$$

Where $\bar{x}$ is the sample mean
  n is the sample size

2.2.5 The sample standard deviation

The **sample standard deviation**, s, is the square root of the sample estimate of the population variance, i.e.:

$$s = \sqrt{\frac{\sum (x_i - \bar{x})^2}{n - 1}}$$

Where $\bar{x}$ is the sample mean
  n is the sample size

2.2.6 Degrees of freedom

The value $n - 1$ in the above equations for variance and standard deviation is referred to as the **degrees of freedom (df)**. This important concept will be met many times in the following chapters and is explained in Box 2.1.

2.2.7 Coefficient of variation

It can be helpful when comparing the amount of variability in parameters whose means and standard deviations are very different to express the standard deviations as a percentage of the respective mean. This statistic is called the **Coefficient of variation (CV)**, that is:

$$CV = \frac{s}{\bar{x}} \bullet 100\%$$

To illustrate, consider the following example. Paediatricians recorded the mean weight at age 2 years of a large number of boys as 14.2 kg with a standard deviation of 1.3 kg. Their mean height was recorded as 89.5 cm with a standard deviation of 3.0 cm. Therefore:

Box 2.1: Degrees of freedom

In statistical analysis, we use the observations in our sample to make estimates of unknown characteristics of the larger population from which our sample was taken. The degrees of freedom in our analysis is the number of observations that are allowed to vary if our sample characteristic is to estimate precisely the population characteristic.

For instance, when we are estimating just one population characteristic like the population variance and our sample size is n, the degrees of freedom is $n - 1$ since control of just one observation (i.e. the rest are free to vary) is all that is required to make our sample variance exactly equal to the population variance.

As an example, suppose we have sample of 5 observations (x_1, x_2, x_3, x_4, x_5; $n = 5$) from a population whose mean is 6. Observe in the table below how control of the last observation can make our sample mean exactly equal to the population mean of 6 when the first 4 observations are free to change their values.

x_1	x_2	x_3	x_4	(Make x_5)	Sample mean
2	4	7	8	(9)	6
4	6	5	7	(8)	6
3	7	4	9	(7)	6

Here we see that a group of 5 observations being used to estimate a single population characteristic has 4 degrees of freedom. In general, when k population characteristics are being estimated from n observations, the degrees of freedom of the analysis is $n - k$.

The *coefficient of variation of weight* at age 2 was
1.3/14.2 × 100% = 9.2%

The *coefficient of variation of height* at age 2 was
3.0/89.5 × 100% = 3.4%.

We can say that the height of boys at this age varies much more than their weight.

2.2.8 The standard error of the mean

The **standard error of the mean** (sem) is the theoretical standard deviation of *all* sample means of size *n* that could be drawn from a population. Its value depends on both the population variance (σ^2) and the sample size (*n*):

$$sem = \frac{\sigma}{\sqrt{n}}$$

where σ^2 is the population variance
   n is the sample size

Since in general we don't know the population variance, the best estimate for the standard error of the mean is calculated as:

$$sem = \frac{s}{\sqrt{n}}$$

where *s* is the sample standard deviation
   n is the sample size

2.2.9 Confidence interval for the mean

The properties of the normal curve (see Chapter 1: *Basic concepts of statistics*) apply to the distribution of sample means. As we previously discussed, if we define the total area under the normal curve as 100%, then the area under

Box 2.2: Using the sem vs the standard deviation in graphical presentations

Some scientists object to the use of the sem in graphical presentations because they feel that the presenter may be understating the variability in their data and that the standard deviation should be used instead. In fact, the appropriate use of the sem or the standard deviation really depends on the context of the presentation.

It is important to remember that the standard deviation is a *sample* property and the sem is an estimate of a *population* property. Therefore, in presentations where several samples are presented and inferences made about their respective population, it is more appropriate to use the sem as the measure of variability.

the curve from one standard deviation below the population mean to one standard deviation above the population mean ($\mu - s$ to $\mu + s$) is 68.26% of the total area. Therefore 68.26% of a large number of estimates of sample mean would lie within ±1 standard error of the mean. In other words, we can be about 68% confident that the population mean lies within ±1 standard error of the sample mean that we measure. However 68% does not provide us with a great deal of certainty. In practice we prefer to state the 95% or 99% **confidence interval**. For large samples ($n > 30$) the appropriate values are ±1.96 sem and ±2.58 sem, respectively The values of 1.96 and 2.58 are the z-scores we used to describe the properties of the normal curve. These values rely on our measurements of $\bar{x}$ and s being good estimates of the population mean and standard deviation. For smaller values of n, the sample standard deviation, s, is a less reliable estimate of population standard deviation, s. We need to increase therefore the values of 1.96 and 2.58 to new values defined by a distribution called **Student's t-distribution** where the appropriate values are dependent on n. Thus for small samples we look up the tabulated critical value of t for the appropriate number ($n - 1$) of **degrees of freedom**; in this case $d.f. = n - 1$. Then the 95% confidence interval $= \bar{x} \pm (t_{0.05} * sem)$. Generally, the confidence interval (**c.i**) is given as:

$$c.i._\alpha = \bar{x} \pm (t_\alpha \bullet sem)$$

where α is the probability level

We shall meet the t-distribution again in Chapter 4.

 ## Step-by-step tutorial 2.2: measures of variability

An agronomist wanted a general idea of the size of potatoes in a crop. He collected 12 potatoes at random locations in the field and recorded their weights in grams as:

230, 225, 215, 245, 222, 216, 245, 228, 255, 237, 199, 207

The **mean** $\bar{x}$ = (230 + 225 + 215 + 245 + 222 + 216 + 245
$\qquad\qquad$ + 228 + 255 + 237 + 199 + 207)/12
$\qquad\qquad$ = 2724/12 g
$\qquad\qquad$ = 227 g

Step 1 For the computation of the sample variance, we tabulate the data as in Table 2.1.

Step 2 From Table 2.1, the sample **variance** is calculated as:

$$s^2 = \frac{\sum(x_i - \bar{x})^2}{n - 1}$$
$$= 3020/11$$
$$= 274.54$$

(A simpler way to calculate $\Sigma(x_i - \bar{x})^2$ is shown in Box 2.3, p. 12.)

Table 2.1 Computing the sample variance

x_i	$x_i - \bar{x}$	$(x_i - \bar{x})^2$
230	3	9
225	−2	4
215	−12	144
245	18	324
222	−5	25
216	−11	121
245	18	324
228	1	1
255	28	784
237	10	100
199	−28	784
207	−20	400

$$\sum (x_i - \bar{x})^2 = 3020$$

Step 3 The sample **standard deviation** is calculated as:

$$s = \sqrt{\frac{\sum (x_i - \bar{x})^2}{n-1}}$$

$$= \sqrt{\frac{3020}{11}}$$

$$= 16.57\,g$$

Step 4 The **standard error of the mean** is calculated as:

$$sem = \frac{s}{\sqrt{n}}$$

$$= 16.57/3.46\,g$$

$$= 4.79\,g$$

Step 5 The 95% **confidence interval** is calculated as follows:

Degrees of freedom $= n - 1 = 12 - 1 = 11$

From the t-tables given in Appendix 1, the critical value of t for 11 degrees of freedom at the $\alpha = 0.05$ level is 2.201. Hence the 95% confidence interval (c.i.) is given by:

$$c.i. = \bar{x} \pm 2.201 \times sem$$

$$= 227 \pm 2.201 \times 4.78\,g$$

$$= 227 \pm 10.52\,g$$

Thus we can be 95% confident that the population mean weight lies between 237.52 and 216.48 g.

2.3 Shape of the distribution: normality

An important aspect of the description of a variable is the *shape of its distribution*. For many of the tests to be described later we need to know how well the variable's distribution can be approximated by the normal distribution (see Chapter 1: *Basic concepts of statistics*). Descriptive statistics provides 2 parameters which can help to assess the normality of your data. These are **skewness** and **kurtosis**.

2.3.1 Skewness

Skewness is a measure of symmetry. A distribution is **symmetric** if it looks similar both sides of the centre. The histogram in Figure 2.1 shows the response times in milliseconds of a subject in a psychological test.

There is visual evidence in Figure 2.1 of a distribution that is skewed to the right – i.e. it was more common for the subject to have a longer rather than shorter than average response time.

Skewness is calculated as follows:

$$Skewness = \frac{n}{(n-1)(n-2)} \bullet \frac{\sum_{i-1}^{n}(x_i - \bar{x})^3}{s^3}$$

where x_i is the value of the ith observation, $\bar{x}$ is the sample mean, s is the sample standard deviation and n is the number of observations.

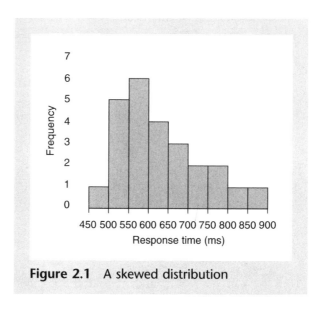

Figure 2.1 A skewed distribution

For the normal distribution the skewness is zero – i.e. the distribution is perfectly symmetrical. Negative values for the skewness indicate data that are skewed to the left (**negatively skewed**) and positive values indicate data that are skewed to the right (**positively skewed**).

The standard error of the skewness, SE_{skew}, is given by the formula:

$$SE_{skew=} = \sqrt{\frac{6}{n}}$$

For normally distributed data, $\dfrac{skewness}{SE_{skew}}$ lies between –1.96 and +1.96.

2.3.2 Kurtosis

Kurtosis measures whether the data are *peaked* or *flat* relative to a normal curve (Figure 2.2).

The two distributions above have similar variance and skew but different kurtosis. The distribution on the left with the extended tails has *positive* kurtosis and that on the right *negative* kurtosis.

Kurtosis is calculated as follows:

$$Kurtosis = \frac{n(n+1)}{(n-1)(n-2)} \bullet \frac{\sum_{i=1}^{n}(x-\bar{x})^4}{s^4} - \frac{3(n-1)^2}{(n-2)(n-3)}$$

where x_i is the value of the *i*th observation, $\bar{x}$ is the sample mean, s is the sample standard deviation and n is the number of observations.

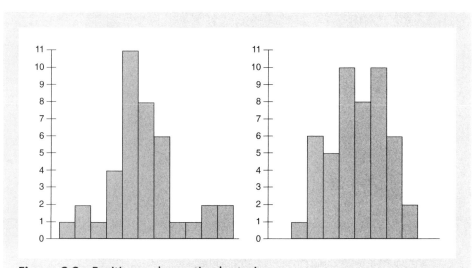

Figure 2.2 Positive and negative kurtosis

The kurtosis of the normal distribution is zero. If the kurtosis is different from zero, then the distribution is either flatter or more peaked than normal. The standard error of the kurtosis, SE_{kurt}, is given by the formula:

$$SE_{kurst} = \sqrt{\frac{24}{n}}$$

For normally distributed data, $\dfrac{kurtosis}{SE_{kurt}}$ lies between -1.96 and $+1.96$.

Hence to check data for normality prior to selecting a statistical test you can calculate the values of $\dfrac{skewness}{SE_{skew}}$ and $\dfrac{kurtosis}{SE_{kurt}}$ and check that both fall between -1.96 and $+1.96$. If either are outside these limits the data are unlikely to be normally distributed.

Other **normality tests** to determine the probability that the sample came from a normally distributed population of observations are also available (e.g. the Kolmogorov–Smirnov test, or the Shapiro–Wilks' W test) but are beyond the scope of this book. None of these tests, however, can substitute for a visual examination of the data using a histogram (see Chapter 3: *Histograms and box plots*). Central classes should have higher frequencies with the peripheral classes having lower frequencies. This is the simplest approach to assessing normality.

Box 2.3: An alternative formulation for the sum of squared deviations

For calculation by hand or calculator of standard deviations the following alternative formulation of $\Sigma(x_i - \bar{x})^2$ provides a simpler calculation:

$$\sum (x_i - \bar{x})^2 = x_1^2 - 2\bar{x} \bullet x_1 + \bar{x}^2 + x_2^2 - 2\bar{x} \bullet x_2 + \bar{x}^2 \ldots + x_n^2 - 2\bar{x} \bullet x_n + \bar{x}$$

$$= \sum x_i^2 - 2\bar{x} \bullet \sum x_i + n \bullet \bar{x}^2$$

$$= \sum x_i^2 - 2\bar{x} \bullet n\bar{x} + n\bar{x}^2 \text{ (since } \sum x_i = n\bar{x})$$

$$= \sum x_i^2 - n\bar{x}^2$$

$$= \sum x_i^2 - n \bullet \left(\frac{\sum x_i}{n}\right)^2$$

$$= \sum x_i^2 - \frac{(\sum x_i)^2}{n}$$

Thus for each value it suffices to store the value and square of the value and sum each of these.

PractiStat tutorial 2.1: Descriptive statistics

This tutorial uses data from Step-by-step tutorial 2.2 (p. 24).

Step 1 Select the OPEN button on the toolbar or select *Open data file* from the *File* menu and navigate to the 'Samples' folder (directory) on the Practi-Stat CD-ROM. Open the data file titled 'Potatoes'. The data from this file should now be displayed and the name 'Potatoes' should appear on the samples list.

Step 2 Select *Descriptive* from the *Statistics* menu or click on the DESC button. The program should now show the descriptive statistics for our sample of weights of potatoes (in grams).

Step 3 The first thirteen boxes ('Mean' through 'Coefficient of Variation') describe measures of central tendency and variability in weights of the sample of 12 potatoes. The last two boxes together give an estimate of the entire crop's mean weight (the population mean) at the 95% level of confidence ($\alpha = 0.05$). In other words, the mean weight of potatoes of the entire crop is *likely* to be between 216 and 238 grams.

3

Histograms and box plots

Using graphic plots to visualise data distributions

Chapter 3 explains how the visual representation of our data helps us to see trends and patterns and then to interpret them. We examine how to give a rough visual impression of the data by a dot plot, how to ascribe continuous data to classes and so construct a histogram and how to represent the data's minimum/maximum values, median and quartiles in a box plot. We then show how to visualise and interpret any outliers or erroneous values in the data.

A visual representation of data can make it easier to spot trends and patterns and to interpret the information. In this chapter we describe the most suitable methods for presenting biological data to visualise distributions in our data.

3.1 Dot plot

A **dot plot** is a simple way of presenting data that gives a rough visual impression of the distribution. A horizontal line is marked with divisions corresponding to the variable being measured. Each observation is entered as a point above the appropriate point on the scale. Repeated observations at the same scale point are stacked on top of each other.

For example, a zoologist recorded the number of live young in the first litter of 70 female mice as shown in Table 3.1.

Table 3.1 Litter sizes in mice

Litter size	1	2	3	4	5	6	7	8	9	10	11	12	13
Frequency (number of mice)	1	2	4	4	6	7	11	10	8	7	5	3	2

A dot plot of these data is shown in Figure 3.1. A horizontal line is marked with the numbers 1 through 13 to represent the litter sizes. Each female mouse is entered as a point above the number on the scale that corresponds to her litter size.

We can easily locate the mode of the distribution (litter size of 7) and see that litter sizes in these type of mice seem to be normally distributed around the mode with fewer mice having very small litters or very large litters.

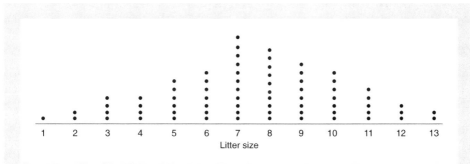

Figure 3.1 Dot plot presentation of the distribution of litter sizes in mice

A limitation to the dot plot is its inability to properly handle **continuous variables**. If we needed to visualise the distribution of measures like 7.1 grams, 7.3 grams, 6.8 grams and 7.2 grams, we would be better served with a **histogram**.

3.2 Histogram

The most usual graphical representation of a sample's distribution is as a histogram. For discrete data such as the litter sizes given above, the histogram is plotted in a straightforward manner as shown in Figure 3.2.

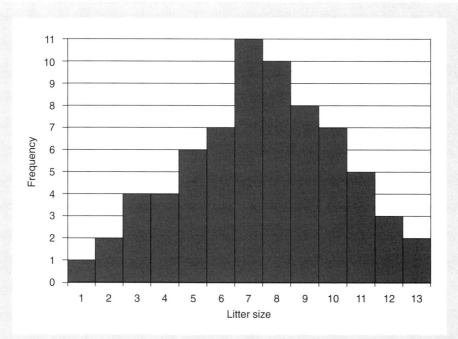

Figure 3.2 Histogram presentation of the distribution of litter sizes in mice

However for continuous variables we must assign the data to chosen **class intervals**. The variable is then plotted in class intervals along the x-axis and the frequency of occurrence of the observations within each interval is plotted on the y-axis. The area of each block on the histogram is proportional to the frequency. If the size of the intervals is constant then the height is also proportional to the frequency. The choice of the optimum number and size of the intervals is important. For example, suppose we record the weight of 500 university students. Assigning the data to 7 classes with a class interval of 10 kg results in the histogram in Figure 3.3.

Notice that whether we express the data as absolute number of observations (left **ordinate** in Figure 3.3) or as the proportion of observations within each class interval – i.e. relative frequency (right ordinate) – the

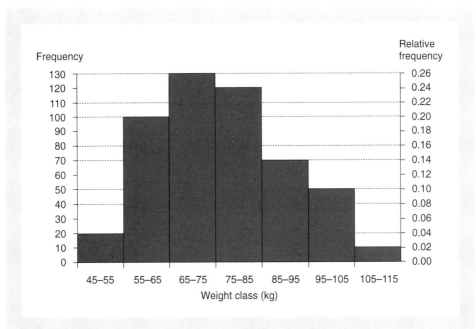

Figure 3.3 Histogram presentation of the distribution of weights of 500 students

shape of the histogram is the same. Note also that continuous intervals may be left- or right-exclusive. For example if the 45–55 continuous interval is right-exclusive then the number 54.9999 falls into the class, but the number 55.0000 is excluded. The number 55.0000 would belong to the 55–65 class interval. The choice of left- or right-exclusivity is really arbitrary since it has little effect on the general shape of the histogram.

The data appear reasonably symmetrical with a peak value in the 65–75 kg class. However when the data are plotted with an increased number (14) of classes and a class interval of 5 kg the histogram in Figure 3.4 is obtained.

The histogram now is seen to be **bimodal**, possibly corresponding to the different weight distributions of men and women.

Although there are no hard and fast rules for deciding on the optimum class interval, a useful rule of thumb is to use $\sqrt{n}$ classes, where n is the total number of observations. It is also important to consider the degree of precision required for the measurements. In the above example weights were recorded to the nearest kg and there was a difference of about 70 kg between the lowest and highest value – i.e. about 70 unit steps. Values from 30 to 300 unit steps between the extreme values are usually regarded as affording adequate precision.

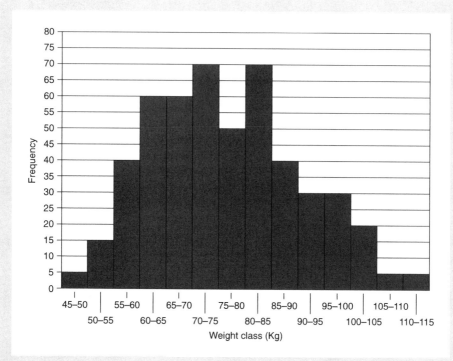

Figure 3.4 Presentation of student weights with an increased number of class intervals

3.3 Box plot

The **box plot** is a very informative way of graphically presenting your data. It is essentially a visual representation of the minimum and maximum values, the median, and the lower and upper quartiles.

The blood pressures of 9 undergraduates were recorded as 120, 114, 134, 127, 119, 108, 122, 114 and 122 mmHg. Using the procedures of Chapter 2 we can calculate the required parameters for a box plot:

Minimum = 108
Maximum = 134
Median = 120
Q1 = 114
Q3 = 123.25
IQR = 123.25 − 114 = 9.25

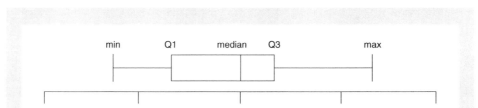

Figure 3.5 Box plot showing minimum, maximum, median and quartile values

To construct a box plot, first draw a line and mark the positions of the minimum, maximum, median and quartile values. A box is then drawn to connect the quartiles and horizontal lines drawn from the minimum to Q1 and from Q3 to the maximum, as shown in Figure 3.5.

The length of the box is the interquartile range (IQR). The centre and spread of the data are clearly indicated and the relative size of the lines from the quartiles to the maximum and minimum values give a visual indication of how skewed the data are. In Figure 3.5 the line from Q3 to the maximum value is almost twice as long as the line from Q1 to the minimum value, showing that these data are indeed skewed.

For normally distributed data, (i) the interval from the minimum to Q1 and from Q3 to the maximum should be approximately of the same size, (ii) the interval from Q1 to the median and from the median to Q3 should be approximately of the same size and (iii) the interval from the minimum to Q1 should generally be larger than that from Q1 to the median.

3.4 Outliers

The box plot is also a useful way to visualise the presence of **outliers** in your data. An outlier is defined as a data point that lies below the *lower* **outlier threshold** or above the *upper* **outlier threshold**. The values at which the thresholds are set is somewhat arbitrary; a common procedure is to use a value of 1.5 times the interquartile range to define the thresholds as follows:

Lower outlier threshold (LOT) $= Q1 - 1.5 \times IQR$

Upper outlier threshold (UOT) $= Q3 - 1.5 \times IQR$

For example, the lower and upper outlier thresholds for the blood pressure data given above are:

$LOT = Q1 - 1.5 \times IQR = 114 - 1.5 \times 9.25 = 100.1$

and

$$UOT = Q3 + 1.5 \times IQR = 123.25 + 1.5 \times 9.25 = 137.1$$

If any data are determined to be outliers then they should be indicated on the box plot as single points (e.g. asterisks or bullets) and the lines from the quartiles redrawn to extend to the minimum and maximum values that are not outliers. For the data given above for blood pressure, all the data (108 to 137) lie within the LOT (101.1) and the UOT (137.1); hence the data contain no outliers.

To demonstrate the occurrence of an outlier, let us substitute for the value of 127 a new value of 91 mm Hg. The new data are:

120, 114, 134, 91, 119, 108, 122, 114, 122

Calculated parameters are:

Minimum = 91
Maximum = 134
Median = 119
Q1   = 112.5
Q3   = 122
LOT   = 98.25
UOT   = 136.25
IQR   = 122 − 112.5 = 9.5

The value 91 is below the LOT and therefore constitutes an outlier. We then draw the box plot as in Figure 3.6.

The asterisk to the left indicates the value of 91 as an outlier and the line from the first quartile (Q1) was redrawn to extend to the lowest value (108) greater than the LOT (98.25).

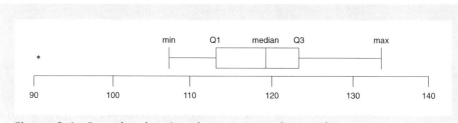

Figure 3.6 Box plot showing the presence of an outlier

3.5 How to handle outliers

It may be tempting to discard outliers when we perceive them to be too extreme to be 'real'. In fact, extreme observations can be an indication that something unusual (and most likely interesting) is being observed. For example, an extreme measure taken from a patient may indicate some major difference in that individual such as an undiagnosed disease or illness.

If there is evidence or logical reason to believe that an observation was incorrectly recorded or measured (e.g. the weight of a 5′ 10″ male recorded as 2,154.345 kg), then that datum's exclusion may be justified.

There are even some instances where incorrectly measured data should not be discarded, such as when the cause of the erroneous data lies within the scope of the stated hypothesis. For example, in a study to determine if the *presence of loud noises* affected the accuracy of temperature readings by hospital staff, the temperatures recorded by 6 nurses on the same patient within a 15-minute period were 36.8, 36.9, 37.3, 45.6, 37.0 and 37.1°C. Although the '45.6' measure was clearly erroneous, that datum should not be discarded since the error may have been a result of the *presence of loud noises*.

When we cannot justify the exclusion of outliers in our data, statistical methods that are immune to the presence of extreme data should be used. Box 1.1 in Chapter 1 (*Which test to use?*) shows the pathways of statistical analyses for such data.

Step-by-step tutorial 3.1: graphical representation of data

A paediatrician investigating the factors affecting infant birth weight recorded the birth weight in kilograms of 50 babies born to non-smoking mothers. The data obtained were as follows in Table 3.2.

Table 3.2 Birth weights of 50 babies born to non-smoking mothers (kg)

3.83	3.38	3.69	2.38	3.15
3.33	3.29	3.43	3.43	3.87
3.55	2.58	3.48	4.16	3.54
3.83	3.21	3.12	3.63	3.28
3.60	3.52	3.43	3.41	2.92
2.21	3.75	3.78	3.40	2.41
1.75	3.52	3.35	3.50	3.72
2.85	2.90	3.03	3.74	3.00
3.17	2.64	2.18	3.37	2.95
3.52	3.92	3.30	2.12	3.61

Step 1   The lowest weight is 1.75 kg and the highest is 4.16. The range is therefore 4.16 − 1.75 = 2.41 kg. The unit step for measurement is 0.01 kg. The range is therefore 241 times the unit step and the data may be regarded as having adequate precision.

Table 3.3 Birth weights: placing data in classes

Class (kg)	Weights (kg)	Frequency
1.60–2.00	1.75	1
2.00–2.40	2.21, 2.18, 2.38, 2.12	4
2.40–2.80	2.58, 2.64, 2.41	3
2.80–3.20	2.85, 3.17, 2.90, 3.12, 3.03, 3.15, 2.92, 3.00, 2.95	9
3.20–3.60	3.33, 3.55, 3.52, 3.34, 3.29, 3.21, 3.52, 3.52, 3.43, 3.48, 3.42, 3.35, 3.30, 3.43, 3.41, 3.40, 3.50, 3.37, 3.54, 3.28	20
3.60–4.00	3.84, 3.84, 3.60, 3.75, 3.92, 3.69, 3.78, 3.63, 3.74, 3.89, 3.72, 3.61	12
4.00–4.40	4.16	1

Step 2 For representing the data as a histogram, the $\sqrt{n}$ rule of thumb suggests that we take 7 classes. A convenient class interval would be 0.4 kg. The data can be placed into these classes, as shown in Table 3.3.

Note that we chose to make our class intervals right-exclusive and the value of 3.60 kg was excluded from the 3.20–3.60 class interval and included in the 3.60–4.00 interval.

Step 3 The frequencies of the weight classes are then plotted as a histogram, as shown in Figure 3.7.

Step 4 For presentation of the data as a box plot first calculate the required parameters as described in Chapter 2:

Minimum = 1.75
Maximum = 4.16
Median = 3.39
Q1   = 3.00
Q3   = 3.60
IQR   = 3.60 − 3.00 = 0.60

Step 5 Calculate the outlier threshold values:

Lower outlier threshold $(LOT) = Q1 - 1.5 \times IQR = 2.10$

Upper outlier threshold $(UOT) = Q3 - 1.5 \times IQR = 4.50$

Since our minimum value of 1.75 is an outlier, we adjust the box plot minimum to the lowest value that is not an outlier:

Box plot minimum = 2.12

Step 6 Draw the box plot as shown in Figure 3.8.

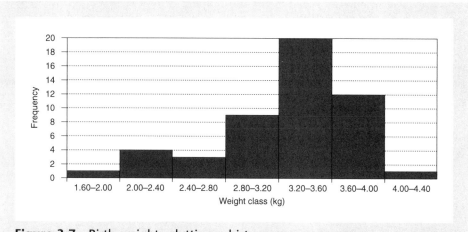

Figure 3.7 Birth weight: plotting a histogram

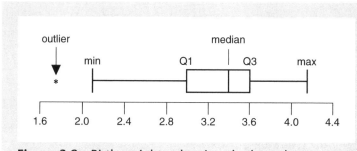

Figure 3.8 Birth weights: drawing the box plot

 PractiStat tutorial 3.1: graphical representation of data

PractiStat provides a convenient and simple means of graphically showing the distribution of observation in our sample by means of a histogram or box plot. This tutorial uses data from Step-by-step tutorial 3.1.

Step 1   Select the OPEN button on the toolbar or select *Open data file* from the *File* menu and navigate to the "Samples" folder on the PractiStat CD-ROM. Open the data file "Birth Weights-Kg". The data from this file should now be displayed and the name "Birth Weights-Kg" should appear on the samples list.

Step 2   Select *Box Plot* from the *Statistics* menu, or click on the BOXPLOT button. The viewing area should now show a box plot and list some important box plot parameters like the minimum, maximum, first and third quartiles, the outlier thresholds and the number of outliers present in the sample. The minimum value of 1.75 kg is an outlier and is presented as a 'bullet' on the box plot. Note that the vertical line that indicates the minimum has been redrawn at 2.12 kg to indicate the minimum value that is not an outlier. The box plot gives us a quick visual distribution of the observations in our sample: the first quartile (1.75 kg–3.00 kg) occupies nearly half of the total sample range while observations in the third quartile are closely packed between 3.39 kg and 3.61 kg.

Step 3 Select *Histogram* from the *Statistics* menu or click on the HIST button. The viewing area should now show a histogram. Press the "Edit Histogram . . ." button and then press the "Suggest . . ." button on the following dialogue. PractiStat suggests 7 classes in the range from 1.509 to 4.401. A class interval magnitude of 0.629 is shown below the upper class limit. To give ourselves a convenient class interval of 0.40, change the Lower Class Limit to 1.60 and the Upper Class Limit to 4.40. Now press the "OK" button.

Step 4 To demonstrate the effects of your choice of histogram plot parameters, use the "Edit Histogram . . ." button to make parameter changes and see what the resulting histogram plots look like. When the "Suggest . . ." button is selected, *PractiStat* sets the parameters as follows:

Number of classes = $\sqrt{n}$

Lower limit = x_{min} less 10% of the data range

Upper limit = x_{max} plus 10% of the data range

PractiStat will also restrict your selection of parameter values according to the following constraints:

Minimum number of classes = 1

Lower class limit ≤ x_{min}

Upper class limit > x_{max}

4

Student's t-test

Comparing two samples

Chapter 4 examines how the important unpaired or independent *t*-test known as the 2-tailed Student's *t*-test can be use to determine whether the means of two independent samples are different enough for us to conclude that they were drawn from different populations. We first show how to use *t*-tables to conclude that there is or is not statistical evidence that there is significant difference in our results from each of our two sets of experimental conditions. We next look at how the variance ratio test (Snedecor's *F*-test) and the approximate *t*-test can be used to compare two estimates of the value of our population variance, and finally determine whether a 1-tailed test is more appropriate for our data.

4.1 Description

Student's *t*-test, sometimes referred to as an *unpaired* or *independent t*-test, is used to determine-whether the means of two independent samples are different enough to conclude that they were drawn from *different populations*.

4.2 Uses

Student's *t*-test is useful for testing for significant differences between results obtained under two experimental conditions. First, we hypothesise that the means of our two populations are not different (the **null hypothesis**). We then determine the probability (our P-value) that the difference in our samples' means could have arisen by chance. This P-value is thus a measure of the compatibility between our experimental observations and our null hypothesis. A low P-value, say <0.05, is typically regarded as statistical evidence to reject the null hypothesis and conclude that there is significant difference in the results obtained from the two experimental conditions.

4.3 Formulae

For 2 samples of n_1 and n_2 number of observations and with means and standard deviations of $\bar{x}_1$, s_1 and $\bar{x}_2$, s_2, respectively:

$$t = \frac{|\bar{x}_1 - \bar{x}_2|}{\sqrt{\left[\dfrac{(n_1-1)s_1^2 + (n_2-1)s_2^2}{n_1+n_2-2}\right] \bullet \left[\dfrac{1}{n_2}+\dfrac{1}{n_2}\right]}}$$

Degrees of freedom: $df = n_1 + n_2 - 2$

4.4 Use of *t*-table (Appendix 1)

First, we find the critical value of t ($t_{critical}(\alpha; df)$, where α is our probability level for acceptance of significant difference (α is typically set at 0.05 by most researchers in the biological sciences) and *df* is our degrees of freedom. If $t \geq t_{critical}$ then we reject the null hypothesis and accept the alternative hypothesis that our two experimental groups produced significantly different results.

 Step-by-step tutorial 4.1: 2-tailed *t*-test

Angioplasty is a surgical procedure that uses a catheter and a balloon to open up blocked arteries, usually those of the heart. A cardiologist wanted

Table 4.1 Angioplasty balloon burst pressures (bar)

Supplier 1 (current)					Supplier 2 (new)				
18.4	20.8	18.0	14.6	18.8	16.2	18.4	17.6	17.8	17.8

to verify that a second supplier of angioplasty balloons was producing balloons of equivalent strength as the current supplier. Five balloons from each supplier were randomly selected and tested for their burst pressures, measured in bars (1 bar = $1\,kg.cm^{-2}$). The data obtained were as shown in Table 4.1.

Step 1 Calculate the means and standard deviations of the two samples (see Chapter 2: *Descriptive Statistics*):

> **Mean$_1$** = 18.12
> **SD$_1$** = 2.24

> **Mean$_2$** = 17.56
> **SD$_2$** = 0.82

Step 2 Calculate the absolute difference in means, $\Delta\bar{x}$:

$$\Delta\bar{x} = |\text{Mean}_1 - \text{Mean}_2| = |18.12 - 17.56| = 0.56$$

Step 3 Calculate the quantity, $k_1 = (n_1 - 1) \times s_1^2 + (n_2 - 1) \times s_2^2$:

$$k_1 = \left((5-1) \times (2.24)^2\right) + \left((5-1) \times (0.82)^2\right) = 22.76$$

Step 4 Calculate the degrees of freedom, df:

$$df = n_1 + n_2 - 2 = 5 + 5 - 2 = 8$$

Step 5 Calculate the quantity $k_2 = \dfrac{1}{n_1} + \dfrac{1}{n_2}$:

$$k_2 = \frac{1}{5} + \frac{1}{5} = 0.4$$

Step 6 Calculate t:

$$t = \frac{\Delta\bar{x}}{\sqrt{\dfrac{k_1}{df} \bullet k_2}} = \frac{0.56}{\sqrt{\dfrac{22.76}{8} \bullet 0.4}} = 0.52$$

From our tables of t (Appendix 1), for a 2-tailed test at an acceptance level of $\alpha = 0.05$, we find that the critical value of t at 8 degrees of freedom is 2.306. Our calculated value of t, 0.52, is less than this critical value, so we therefore do not reject the null hypothesis. Our data do not provide sufficient evidence that the observed difference in the mean values of our two samples has arisen other than by chance. We can therefore conclude that the new supplier is producing angioplasty balloons that are equivalent in strength to the current supplier.

 PractiStat tutorial 4.1: 2-tailed *t*-test

Step 1 Select the *Open data file* menu item from the *File* menu or click on the OPEN button and navigate to the 'Samples' folder on the PractiStat CD-ROM. Open the file titled 'Supplier Current'. On your computer screen, the data from this file should now be displayed and the name 'Supplier Current' should appear on the samples list.

Step 2 Similarly open the file titled 'Supplier New'. The data from this file should now be displayed and the name 'Supplier New' should appear on the samples list.

Select Student's t-test from the *Statistics* menu or click the T-TEST button on the toolbar and select the '2-tailed' option in the 'Options' box at the upper right of the PractiStat Window. Multi-select 'Supplier Current' and 'Supplier New' from the samples list (Control-click on PCs or Command-click on Macs).

The viewing area should now show the results of a 2-tailed Student's *t*-test. The calculated *t*-value is 0.524 and *t*-critical with 8 degrees of freedom at the 0.05 significance level is 2.306. The P-value is listed as non-significant ('NS (P > 0.05)'). Therefore we can conclude that the new supplier is producing angioplasty balloons that are equivalent in strength to the current supplier.

 4.5 Assumptions underlying the *t*-test

When performing Student's *t*-test, the following 2 assumptions are made:

1 The 2 samples were randomly drawn from normally distributed populations

2 The variances of the two populations are identical.

Therefore before carrying out the *t*-test we should test the samples for their skewness, the presence of outliers (see Chapter 3: *Histograms and box plots*), and for equality of variance using the variance ratio test (**Snedecor's *F*-test**) shown in **4.6** below.

Box 4.1 What student?

W. S. Gossett was working for the Guinness Brewing Company in Dublin at the time he developed the *t*-distribution and related *t*-tests. Although as a rule the brewery forbade its employees from publishing company research, it allowed Gossett to publish his famous work in 1908 under the pseudonym 'Student'.

If any of the samples have outliers or if there is reason to believe that they were drawn from skewed populations, then the non-parametric **Mann–Whitney U-test** may be more appropriate (see Chapter 5: *The Mann–Whitney U-test*).

If both samples were drawn from normally distributed populations and the Snedecor's *F*-test fails to show equality of variance, then the approximate *t*-test may be used (see below).

4.6 Variance ratio test (Snedecor's *F*-test)

The variance ratio test (*F*-test) compares two estimates of the value of the population variance:

$$F = \frac{\text{Greater estimate of population variance}}{\text{Lesser estimate of population variance}}$$

or

$$F = \frac{s_{greater}^2}{s_{lesser}^2}$$

$$df_{numerator} = df_{greater} = (n_{greater} - 1)$$

$$df_{denominator} = df_{lesser} = (n_{lesser} - 1)$$

The two estimates are those derived from the two samples – i.e. s_1^2 and s_2^2. Using the table of *F* at the 5 per cent level (Appendix 2, p. 50), you should locate the column for the number of degrees of freedom of the sample giving the larger estimate of variance and then the row for the number of degrees of freedom of the sample giving the smaller estimate. Where these intersect is the theoretical value of *F* for the 2 samples. If your calculated value of *F* is greater than the theoretical value then the variances of the 2 samples are significantly different. In this case you may use the 'approximate *t*-test'.

4.7 Approximate *t*-test

When the variances of the 2 samples are significantly different, it is no longer appropriate to combine s_1 and s_2 to obtain a pooled estimate of the population standard deviation (see Box 4.2, p. 47). Instead the following formula is used to calculate '*t*':

$$t = \frac{|\bar{x}_1 - \bar{x}_2|}{\sqrt{\dfrac{s_1^2}{n_1} + \dfrac{s_2^2}{n_2}}}$$

$$df = n_1 + n_2 - 2$$

Step-by-step tutorial 4.2: variance ratio (F)-test and approximate t-test

Measurements were made of the rate of insulin release from islets of Langerhans incubated *in vitro* in the presence of 3.3 mM glucose and in the absence or presence of the sulphonylurea drug tolbutamide. The aim of the experiment was to determine whether tolbutamide was able to alter insulin release at this glucose concentration. The data obtained were as shown in Table 4.2.

Table 4.2 Insulin release (μU·islet^{-1}·h^{-1})

Sample 1 3.3 mM glucose					Sample 2 3.3 mM glucose + 0.1 mM tolbutamide				
12.0	15.0	13.4	16.7	20.2	30.0	25.2	40.1	60.0	50.0

Step 1 Calculate the means and standard deviations of the two samples (see Chapter 2: *Descriptive statistics*):

Mean$_1$ = 15.4
SD$_1$ = 3.1
df$_1$ = 4

Mean$_2$ = 41.1
SD$_2$ = 14.3
df$_2$ = 4

Step 2 Calculate the ratio of the variances (with the larger variance as the numerator):

$$Variance\ ratio = \frac{(SD_2)^2}{(SD_1)^2} = \left(\frac{SD_2}{SD_1}\right)^2 = \left(\frac{14.3}{3.1}\right)^2 = 21.3$$

From our tables of F (Appendix 2), at a significance level of $\alpha = 0.05$, we find that the critical value of F with numerator degrees of freedom of 4 and denominator degrees of freedom of 4 is 9.6; since the calculated value exceeds the critical value, the two variances are significantly different and the approximate t-test must therefore be used.

Step 3 Calculate the absolute difference in means, $\Delta\bar{x}$:

$$\Delta\bar{x} = |\text{Mean}_1 - \text{Mean}_2| = |15.4 - 41.1| = 25.7$$

Step 4 Calculate the quantity, $k = \dfrac{s_1^{\,2}}{n_1} + \dfrac{s_2^{\,2}}{n_2}$:

$$k = \frac{3.1^2}{5} + \frac{14.3^2}{5} = 42.8$$

Step 5 Calculate the degrees of freedom, *df*:

$$df = n_1 + n_2 - 2 = 5 + 5 - 2 = 8$$

Step 6 Calculate *t*:

$$t = \frac{\Delta \bar{x}}{\sqrt{k}} = \frac{25.7}{\sqrt{42.8}} = 3.928$$

From our tables of *t* (Appendix 1), at an acceptance level of $\alpha = 0.05$, we find that the critical value of *t* at 8 degrees of freedom is 2.306. Our calculated value of *t*, 3.928, is greater than this critical value, so we can therefore reject the null hypothesis and conclude that there is a significant difference between the results obtained from our two experimental conditions. We can conclude that tolbutamide significantly stimulates insulin secretion in the presence of 3.3 mM glucose.

4.8 1-tailed and 2-tailed *t*-tests

In the worked examples given above the null hypothesis was that the 2 population means were equal, i.e. $\mu_1 = \mu_2$, and the *t*-test was used to test against the alternative hypothesis that $\mu_1 \neq \mu_2$. This is a so-called **2-tailed** ***t*-test**. It is sometimes the case that *before* the data are collected there is only one reasonable way in which the 2 means could differ, and the alternative hypothesis would be for example $\mu_1 > \mu_2$. It is then appropriate to carry out a **1-tailed** ***t*-test**. First check that the means do indeed differ in the direction postulated. If not, reject the alternative hypothesis. If the means do differ in the right direction, then calculate *t* as above but assess the significance using tables from the critical values for a 1-tailed *t*-test. It is important to stress that the directional alternative hypothesis must be stated before seeing the data. Readers interested in the mathematical basis for Student's *t*-test should consult Box 4.2.

Box 4.2 More on the mathematics of Student's t-test

Suppose we wish to test whether a sample of *n* observations whose mean value is $\bar{x}$ could have come from a population with mean value μ and standard deviation *s*. We calculate the ratio:

$$t = \frac{\text{Difference in means}}{\text{Standard error of the difference between means}} = \frac{|\mu - \bar{x}|}{\sigma/\sqrt{n}}$$

which is called Student's *t*. Since in general we don't know the population standard deviation we must estimate it from the sample standard deviation, *s*:

Hence $t = \dfrac{|\mu - \bar{x}|}{s/\sqrt{n}}$

Continued

Box 4.2 *Continued*

The larger the discrepancy between the sample and population means, the greater is the value of t. Since t has a calculable probability distribution, any particular value of t will be exceeded with a known probability. There will in general be a family of t curves, one for each *degree of freedom*, $n-1$. These t curves are, in fact, quite similar to the standard normal curve and for 30 or more degrees of freedom t-values differ little from corresponding z-values of the normal curve.

The discussion so far has assumed that we wish to compare a single sample with a population. More frequently we want to compare two sets of sample data in order to determine whether any difference between the mean values of the two samples is statistically significant.

Suppose we have two groups with n_1 and n_2 observations and with corresponding means and standard deviations of $\bar{x}_1$, s_1 and $\bar{x}_2$, s_2.

We can combine s_1 and s_2 to make a best estimate of the population standard deviation, σ, as

$$s = \sqrt{\frac{(n_1-1)s_1^2 + (n_2-1)s_2^2}{n_1+n_2-2}}$$

The best estimate of the standard error of the difference of two sample means (SED) is the square root of the sum of the squared individual sample standard errors:

i.e. $SED = s\sqrt{\dfrac{1}{n_1} + \dfrac{1}{n_2}}$

Hence:

$$t = \frac{\text{Difference in means}}{\text{Standard error of the difference between means}}$$

$$= \frac{|\bar{x}_1 - \bar{x}_2|}{\sqrt{\left[\dfrac{(n_1-1)s_1^2 + (n-1)s_2^2}{n+n-2}\right] \cdot \left[\dfrac{1}{n_1} + \dfrac{1}{n_2}\right]}}$$

with $n_1 + n_2 - 2$ degrees of freedom.

Note that when the to population variances are significantly different it is then not appropriate to combine s_1 and s_2 to obtain a pooled estimate of the population standard deviation. instead the SED is calculated as:

$$SED = \sqrt{\frac{s_1^2}{n_1} + \frac{s_2^2}{n_2}}$$

Hence $t = \dfrac{|\bar{x}_1 + x_2|}{\sqrt{\dfrac{s_1^2}{n_1} + \dfrac{s_2^2}{n_2}}}$

5

The Mann–Whitney *U*-test

Comparing two samples using a non-parametric method

Chapter 5 explains how the non-parametric Mann–Whitney *U*-test can be used to determine whether the medians of two independent samples are different enough to conclude that they were drawn from different populations. We first examine the formulae for determining *U* and calculating the P-value (probability), and then demonstrate the experimental use of the *U*-test statistic.

5.1 Description

The Mann–Whitney U-test, a **non-parametric** or **distribution free test**, is used to determine whether the medians of two *independent* samples are different enough to conclude that they were drawn from different populations.

5.2 Uses

The Mann–Whitney *U*-test, like Student's *t*-test, is useful for testing for significant differences between results obtained under two experimental conditions. In contrast to Student's *t*-test however, the Mann–Whitney *U*-test does not make any assumptions about the *nature of the population distribution* and is appropriate for *ordinal* data or for samples that contain **outliers** (see Chapter 3: *Histograms and box plots*).

First, we hypothesise that the data from both samples were drawn from the same population (the **null hypothesis**). We then determine the probability (our P-value) that the observed difference in our samples could have arisen by chance. A low P-value, say <0.05, is often regarded as statistical evidence to reject the null hypothesis and conclude that there is significant difference in the results obtained from our two experimental conditions.

5.3 Formulae

$$U = \text{larger of } U_1 \text{ and } U_2$$

$$U_1 = n_1 n_2 + \frac{n_1(n_1 + 1)}{2} - R_1$$

$$U_2 = n_1 n_2 + \frac{n_2(n_2 + 1)}{2} - R_2$$

Where sample 2 is the larger sample and R_1 and R_2 are the sums of the ranks of the observations in the samples after the observations of both samples are pooled and ranked in ascending order (lowest value is assigned a rank of 1)

5.3.1 Two methods of calculating the P-value

The determination of the P-value using the table of critical values for U in Appendix 3 is limited to samples with 20 or fewer observations. This is because it is impractical to list all of the possible scenarios of sample pairs. Fortunately, the distribution of rank sums approximates the normal distribution when the sample sizes are larger than 20 and so the probability can alternatively be calculated using the *z*-distribution:

$$z = \frac{U - \mu_U}{s_U} = \frac{U - \dfrac{n_1 n_2}{2}}{\sqrt{\dfrac{n_1 n_2 (n_1 + n_2 + 1)}{12}}}$$

5.3.2 Correction for ties

When you use the z-distribution and there are many ties in the data (this occurs frequently in **discrete** or **ordinal variables**), the standard deviation of U, s_u, needs to be reduced by an amount that is dependent on the number of ties in the pooled data:

$$s_U = \sqrt{\frac{n_1 n_2 (n_1 + n_2 + 1)}{12} - \frac{n_1 n_2 \sum [(T_i - 1) T_i (T_i + 1)]}{12(n_1 + n_2)\left((n_1 + n_2)^2 - 1\right)}}$$

Where T_i is the number of ties in the ith set of ties

Note that for a small number of ties, the denominator of the reducing term dominates and the correction to s_u is negligible.

5.4 Use of the U-table (Appendix 3)

First, we find $U_{critical}(\alpha; n_1; n_2)$, where α is our probability level for acceptance of significant difference (α is typically set at 0.05 by most researchers in the biological sciences) and n_1 and n_2 are our respective smaller and larger sample sizes. If $U \geq U_{critical}$ then we reject the null hypothesis and accept the alternative hypothesis that our two experimental groups produced significantly different results.

5.5 Use of the z-distribution

In a 2-tailed test, the values of $z_{critical}(\alpha)$, where α is our probability level for acceptance of significant difference are 1.96, 2.58 and 3.29 for $\alpha = 0.05$, 0.01, and 0.001, respectively. If $z \geq z_{critical}$ then we reject the null hypothesis and accept the alternative hypothesis that our 2 experimental groups produced significantly different results.

Step-by-step tutorial 5.1: Mann–Whitney U-test for samples with n less than 20

In an experiment to compare the effectiveness of administering insulin via a needle and syringe with an alternative method using a needleless administration system, a fixed dose of insulin was administered to rats by the 2 methods and the blood glucose levels (mM) were determined 30 minutes later. The results were as in Table 5.1.

Table 5.1 Administering insulin

Syringe and needle	3.5	1.4	1.8	4.0	4.3	3.4	3.2	4.2	4.8	1.0
Needleless device	4.2	3.5	5.5	3.3	4.1	3.0	2.9	4.0	3.6	

Table 5.2 Administering insulin: pooling and ordering the observations

Glucose	1.0	1.4	1.8	2.9	3.0	3.2	3.3	3.4	3.5	3.5	3.6	4.0	4.0	4.1	4.2	4.2	4.3	4.8	5.5

Table 5.3 Administering insulin: ranking the observations

Glucose	1.0	1.4	1.8	2.9	3.0.	3.2	3.3	3.4	3.5	3.5	3.6	4.0	4.0	4.1	4.2	4.2	4.3	4.8	5.5
Rank	1	2	3	4	5	6	7	8	9.5	9.5	11	12.5	12.5	14	15.5	15.5	17	18	19

Step 1 Pool the observations from both groups and arrange them in ascending order. Use a mark such as an underline to keep track of the observations that belong to the smaller group (Table 5.2).

Step 2 Assign them ranks starting with the lowest value being assigned a rank of 1. Assign the average rank to tied ranks. For example, the value '3.5' occurs twice and collectively occupy the 9th and 10th rank. Since they are a tie, they are both assigned the rank of $(9 + 10)/2 = 9.5$ (Table 5.3).

Step 3 For each group, calculate the sum of their assigned ranks. Remember that for the calculation of this statistic, it is important to identify the larger group as Group 2. In this example, the 'Syringe and Needle' group is Group 2:

$$R_1 = 4 + 5 + 7 + 9.5 + 11 + 12.5 + 14 + 15.5 + 19 = 97.5$$

$$R_2 = 1 + 2 + 3 + 6 + 8 + 9.5 + 12.5 + 15.5 + 17 + 18 = 92.5$$

Step 4 Calculate:

$$U_1 = n_1 n_2 + \frac{n_1(n_1 + 1)}{2} - R_1$$
$$= 9 \times 10 + 9 \times 10 \div 2 - 97.5$$
$$= 37.5$$

Step 5 Calculate:

$$U_2 = n_1 n_2 + \frac{n_2(n_2 + 1)}{2} - R_2$$
$$= 9 \times 10 + 10 \times 11 \div 2 - 92.5$$
$$= 52.5$$

Step 6 Check that $U_1 + U_2 = n_1 \times n_2$:

$$37.5 + 52.5 = 90 = 10 \times 9$$

If $(U_1 + U_2)$ is not equal to $(n_1 \times n_2)$ then review steps 1 through 5 for errors.

Step 7 Assign the greater of U_1 and U_2 to U:

$$U = 52.5$$

Step 8 Look up $U_{critical}$ in Appendix 3 and determine the P-value. At a significance level of 0.05 for a 2-tailed test with $n_2 = 10$ and $n_1 = 9$, the critical value of U is 70. Since the calculated value of U for the experimental data is smaller than the critical value, the data do not provide any justification for rejecting the null hypothesis that the sets of data have equal medians. Therefore we conclude that the 2 methods for administering insulin are equally effective.

 ## PractiStat tutorial 5.1: Mann–Whitney U-test for n less than 20

Step 1 Select the OPEN button on the toolbar or select *Open data file* from the *File* menu and navigate to the 'Samples' folder on the PractiStat CD-ROM. Open the file titled 'Glucose Syringe'. On your computer screen, the data from this file should now be displayed and the name 'Glucose Syringe' should appear on the samples list.

Step 2 Once again, select the OPEN button on the toolbar and open the file titled 'Glucose Needleless'. The data from this file should now be displayed and the name 'Glucose Needleless' should appear on the samples list.

Step 3 Select the MANN-U button on the toolbar or select *Mann-Whitney* U-*test* from the *Statistics* menu and check that the '2-tailed' option is selected in the 'Options' area at the upper right of the PractiStat window. Multi-select 'Glucose Syringe' and 'Glucose Needleless' from the samples list (Control-click on PCs or Command-click on Macs).

Step 4 The viewing area should now show the results of a 2-tailed Mann–Whitney U-test. The calculated U-value is 52.5 and the 2-tailed $U_{critical}$ value for $n_2 = 10$ and $n_1 = 9$ at the 0.05 significance level is 70. Since our calculated U-value is less than $U_{critical}$, the P-value is listed as 'NS (P > .05)'. Therefore we can conclude that the two methods of insulin administration are equally effective.

 ## Step-by-step tutorial 5.2: Mann–Whitney U-test for samples with n greater than 20

In a study of the influence of dietary fat on pancreatic β-cell function, 20 rats were fed on a control diet and 25 on a fat-enriched diet. One parameter

Table 5.4 Influence of dietary fat on pancreatic β-cell function

Control diet					Fat-enriched diet				
32.5	35.5	40.0	23.3	38.9	23.9	32.0	19.7	37.0	26.0
44.2	28.4	21.0	35.1	40.0	28.9	22.0	34.6	19.7	26.0
23.5	36.8	20.7	44.2	38.1	35.0	22.1	17.0	27.6	19.7
19.8	38.4	37.0	41.2	24.5	32.2	16.9	24.9	22.9	30.2
					22.8	40.0	15.9	19.6	25.9

Table 5.5 Fat and β-cell function: pooling and ordering the observations

15.9	16.9	17.0	19.6	19.7	19.7	19.7	19.8	20.7	21.0
22.0	22.1	22.8	22.9	23.3	23.5	23.9	24.5	24.9	25.9
26.0	26.0	27.6	28.4	28.9	30.2	32.0	32.2	32.5	34.6
35.0	35.1	35.5	36.8	37.0	37.0	38.1	38.4	38.9	40.0
40.0	40.0	41.2	44.2	44.2					

measured was the pancreatic β-cell activity of the enzyme glucokinase. The results were as in Table 5.4 with the glucokinase activity expressed as μmoles·min^{-1}·g^{-1}.

Since n_2 contains more than 20 observations we can use the z-distribution to assess the significance of U in this example.

Step 1 Pool the observations from both groups and arrange them in ascending order. Use a mark such as an underline to keep track of the observations that belong to the smaller group (Control Diet) (Table 5.5).

Step 2 Assign them ranks starting with the lowest value being assigned a rank of 1. Assign the average rank to tied ranks. For example, the value '19.7' occurs three times and collectively occupy the 5th, 6th and 7th rank. Since they are a tie, all three are assigned the rank of $(5 + 6 + 7)/3 = 6$ (Table 5.6).

Step 3 For each group, calculate the sum of their assigned ranks. Remember that for the calculation of this statistic, is important to identify the larger group as group 2. In this example, the 'Control Diet' group is group 1 and the 'Fat-enriched Diet' group is group 2:

$R_1 = 8 + 9 + 10 + 15 + 16 + 18 + 24 + 29 + 32 + 33 + 34 + 35.5 + 37$
$\quad + 38 + 39 + 41 + 41 + 43 + 44.5 + 44.5$
$\quad = 591.5$

$R_2 = 1 + 2 + 3 + 4 + 6 + 6 + 6 + 11 + 12 + 13 + 14 + 17 + 19 + 20$
$\quad + 21.5 + 21.5 + 23 + 25 + 26 + 27 + 28 + 30 + 31 + 35.5 + 41$
$\quad = 443.5$

Table 5.6 Fat and β-cell function: ranking the data

Value	15.9	16.9	17.0	19.6	19.7	19.7	19.7	19.8	20.7	21.0
Rank	1	2	3	4	6	6	6	8	9	10
Value	22.0	22.1	22.8	22.9	23.3	23.5	23.9	24.5	24.9	25.9
Rank	11	12	13	14	15	16	17	18	19	20
Value	26.0	26.0	27.6	28.4	28.9	30.2	32.0	32.2	32.5	34.6
Rank	21.5	21.5	23	24	25	26	27	28	29	30
Value	35.0	35.1	35.5	36.8	37.0	37.0	38.1	38.4	38.9	40.0
Rank	31	32	33	34	35.5	35.5	37	38	39	41
Value	40.0	40.0	41.2	44.2	44.2					
Rank	41	41	43	44.5	44.5					

Step 4 Calculate:

$$U_1 = n_1 n_2 + \frac{n_1(n_1+1)}{2} - R_1$$
$$= 20 \times 25 + 20 \times 21 \div 2 - 591.5$$
$$= 118.5$$

Step 5 Calculate:

$$U_2 = n_1 n_2 + \frac{n_2(n_2+1)}{2} - R_2$$
$$= 20 \times 25 + 20 \times 26 \div 2 - 443.5$$
$$= 381.5$$

Step 6 Check that $U_1 + U_2 = n_1 \times n_2$:

$$118.5 + 381.5 = 500 = 20 \times 25$$

If $(U_1 + U_2)$ is not equal to $(n_1 \times n_2)$ then review steps 1 through 5 for errors.

Step 7 Assign either of U_1 or U_2 to U:

$$U = 381.5$$

Step 8 Calculate z:

$$z = \frac{U - \dfrac{n_1 n_2}{2}}{\sqrt{\dfrac{n_1 n_2 (n_1 + n_2 + 1)}{12}}}$$
$$= \frac{381.5 - 250}{\sqrt{\dfrac{500 \times 46}{12}}}$$
$$= 3.004$$

Step 9 Compare this value of z with the critical value of z at the desired level of significance. Our calculated value is greater than the critical value of 2.58 for $P < 0.01$. Hence we can reject the null hypothesis and conclude that there is a significant effect of fat feeding on the activity of glucokinase.

Note that if we use U_1 instead of U_2 we come to the same conclusion. In this case we calculate:

$$z = \frac{118.5 - 250}{\sqrt{\dfrac{500 \times 46}{12}}}$$
$$= -3.004$$

We ignore the sign and again conclude a significant result.

 PractiStat tutorial 5.2: Mann–Whitney U-test for n greater than 20

Step 1 Select the OPEN button on the toolbar or select *Open data file* from the *File* menu and navigate to the 'Samples' folder on the PractiStat CD-ROM. Open the file titled 'Diet Control'. On your computer screen, the data from this file should now be displayed and the name 'Diet Control' should appear on the samples list.

Step 2 Once again, select the OPEN button on the toolbar and open the file titled 'Diet Fat-Enriched'. The data from this file should now be displayed and the name 'Diet Fat-Enriched' should appear on the samples list.

Step 3 Select the MANN-U button on the toolbar or select *Mann-Whitney U-test* from the *Statistics* menu and check that the '2-tailed' option is selected in the 'Options' area at the upper right of the PractiStat window. Multi-select 'Diet Control' and 'Diet Fat-Enriched' from the samples list (Control-click on PCs or Command-click on Macs).

Step 4 The viewing area should now show the results of a 2-tailed Mann–Whitney U-test using the z-distribution ($n_2 > 20$). The calculated z-value is 3.004. The critical value of z in a 2-tailed test at the 0.05 significance level is 1.96. Since the calculated z-value exceeds $z_{critical}$, the P-value is listed as '$P < 0.05$'. Therefore we can conclude that there was a significant effect of fat feeding on the glucokinase activity.

 ## 5.6 Assumptions underlying the Mann–Whitney U-test

1 The two samples are drawn randomly and independently.

2 The measures within the two samples are able to be ranked and hence must be continuous, discrete or ordinal. Note that when the data are not

continuous many ties may result. When this occurs z is calculated using a modified formula for the standard deviation of U, s_u, given under the *Formulae* section 5.3 above. Readers interested in the mathematical basis for the Mann–Whitney U-test should consult Box 5.1.

Box 5.1 More on the mathematics of the Mann–Whitney U-test

The sum of the first n integer numbers is given by $\dfrac{n(n+1)}{2}$. Therefore when n numbers are ranked the sum of ranks is $\dfrac{n(n+1)}{2}$.

Suppose that we have two groups with n_1 and n_2 observations in each. Then the **total** sum of ranks is:

$$\frac{(n_1 + n_2)(n_1 + n_2 + 1)}{2}$$

The **maximum value** of the sum of ranks for the group with n_1 observations would occur when the top n_1 rankings are all occupied by members of this group. This maximum sum of ranks for n_1 (R_{1max}) would thus be the (total sum of ranks) − (the sum of ranks for the group with n_2 values):

$$\text{I.e. } R_1 max = \frac{(n_1 + n_2)(n_1 + n_2 + 1)}{2} - \frac{n_2(n_2 + 1)}{2}$$

$$= \frac{n_1^2 + n_1 n_2 + n_1 + n_1 n_2 + n_2^2 + n_2 - n_2^2 - n_2}{2}$$

$$= \frac{2n_1 n_2 + n_1 + n_1^2}{2}$$

$$= n_1 n_2 + \frac{n_1(n_1 + 1)}{2}$$

Thus $U_1 = R_1 max - R_1 = n_1 n_2 + \dfrac{n_1(n_1 + 1)}{2} - R_1$

The minimum value of U_1 is clearly zero, when the top n_1 rankings are all occupied by members of this group. The maximum value of U_1 will be given by $n_1 \cdot n_2$ and will occur when the lowest n_1 rankings are all occupied by members of this group.

Similarly $R_2 max = n_1 n_2 + \dfrac{n_2(n_2 + 1)}{2}$ and $U_2 = R_2 max - R_2 = n_1 n_2 + \dfrac{n_2(n_2 + 1)}{2} - R_2$.

Values of U_2 will range from zero when $U_1 = n_1 \cdot n_2$ to $n_1 \cdot n_2$ when $U_1 = $ zero. Thus U_1 and U_2 are mirror images of each other.

Probability theory tells us that the number of possible ways of ranking two groups of n_1 and n_2 observations is:

$$\frac{(n_1 + n_2)!}{n_1! n_2!}$$

Continued

Box 5.1 *Continued*

For each ranking order the value of U can be calculated as the difference between R_{max} and R. We can then ask what *proportion* of these ranking orders would produce a value of U (the larger of the two values U_1 and U_2) as large as the one we calculate from our data. If this proportion is less than 0.05 we can conclude that there is less than a 5% chance that the observed value of U could have been obtained by mere chance and hence we can say that the result is significant at the 0.05 level. Note that we could also select the smaller of U_1 and U_2 to be U and then ask what proportion of ranking orders would produce a value of U as small as the one we calculate from our data. Tables exist for both these approaches, which of course yield results.

We can illustrate with a simple example. First note that only one ranking order can have a value for U_1 of zero. Therefore for this single order to represent 0.05 of the total number of ranking orders there must be at lest 20 raking orders possible. The lowest number of values of n_1 and n_2 that satisfy this criterion are $n_1 = n_2 = 3$ for which the total number of ranking orders is $\dfrac{6!}{3!3!} = 20$. Therefore for these values of n_1 and n_2 the only significant result is when there is no overlap between the two groups – i.e. the 3 highest ranks are occupied by values from one group. A similar conclusion is reached from the consideration that only one ranking order leads to the maximum value of $U_1 = 3 \times 3 = 9$, again when all the top 3 places in the ranking order are occupied by 1 group. The number of ranking orders that each of the possible values of U have together with the corresponding proportion of the total orders is given in the following table.

U	0	1	2	3	4	5	6	7	8	9
No of ranking orders	1	1	2	3	3	3	3	2	1	1
Proportion	0.05	0.05	0.1	0.15	0.15	0.15	0.15	0.1	0.05	0.05

The distribution is symmetrical about the mean corresponding to a value of U given by $\dfrac{n_1 n_2}{2}$. For $n_2 > 20$ the distribution of U becomes a close approximation to the normal distribution with a population mean μ_u of $\dfrac{n_1 n_2}{2}$ and standard deviation s_u of:

$$\sqrt{\frac{n_1 n_2 (n_1 + n_2 + 1)}{2}}$$

Hence for large sample size we can calculate the normal deviate, z, as:

$$z = \frac{U - \mu_u}{s_u} = \frac{U - \left(\dfrac{n_1 n_2}{2}\right)}{\sqrt{\dfrac{n_1 n_2 (n_1 + n_2 + 1)}{12}}}$$

(The above discussion ignores the occurrence of ties. An equation for calculation of s_u that takes account of ties is given in the text.)

6

The paired *t*-test

Comparing paired samples

Chapter 6 explains how the paired *t*-test can be used when our data consist of matched pairs – when the same subject is tested twice, before and after a treatment, for example. We first examine the advantages of excluding within-group variation from our analysis, and then show how *t* can be calculated. We finally compare the results obtained by our non-paired *t*-test with those we would get if we treated our data as independent observations and carried out an unpaired *t*-test.

6.1 Description

A special version of Student's *t*-test is used when the data consist of *matched pairs*. Such data frequently arise when the same subject is tested twice – e.g. before and after a treatment – and the two observations on each subject thus constitute a matched pair.

6.2 Uses

The *t*-test for **paired samples** has the advantage that much of the variations attributable to the initial individual differences between subjects (the so-called 'within-group' variation) is excluded from the analysis. If the two groups of observations to be compared are obtained from the same sample of subjects who were tested twice (e.g. before and after a treatment), instead of analysing the raw data from each group separately we can look only at the *differences* between the pre-treatment and post-treatment values for each subject. By subtracting the first value from the second for each subject and then analysing only these paired differences we exclude all the variation in our data that results from the differing initial values of individual subjects. Therefore, the *t*-test for paired samples is always more sensitive than the *t*-test for independent samples.

6.3 Formulae

For two paired samples of *n* observations:

If x_i is the *i*th observation of the pre-treatment or 'control' sample, y_i is the *i*th observation of the 'treated' sample and d_i is the difference between the two *i*th observations, i.e. $d_i = x_i - y_i$:

$$t = \frac{\sum\limits_{i=1}^{n} d_i}{\sqrt{\dfrac{n \cdot \sum\limits_{i=1}^{n} d_i^2 - \left(\sum\limits_{i=1}^{n} d_i\right)^2}{n-1}}}$$

Degrees of freedom (df) = n – 1

Step-by-step tutorial 6.1: paired *t*-test

Women with premature menopause are at high risk for blood vessel problems. In an investigation of possible factors involved, a study in women undergoing premature (surgical) menopause measured changes in plasma

levels of lipoprotein A, a lipid fraction that is closely associated with formation of blood clots and thickening of arteries. The lipoprotein A levels
(mg/dl) before and after hysterectomy in 10 patients were as in Table 6.1.

Table 6.1 Changes in plasma levels of
lipoprotein A

Patient	Before treatment	After treatment
1	10.3	13.3
2	6.5	9.6
3	6.1	10.8
4	12.5	12.4
5	8.7	12.7
6	14.2	18.1
7	5.0	8.0
8	18.9	21.7
9	6.2	10.5
10	13.1	17.9

Step 1 Calculate and tabulate the difference and the square of the difference in
score for each patient (Table 6.2).

Table 6.2 Lipoprotein A levels: Calculating/tabulating difference and square of
difference

Patient	Before treatment	After treatment	Difference (d)	(Difference)^2 (d^2)
1	10.3	13.3	3.0	9.0
2	6.5	9.6	3.1	9.61
3	6.1	10.8	4.7	22.09
4	12.5	12.4	0.1	0.01
5	8.7	12.7	4.0	16.0
6	14.2	18.1	3.9	15.21
7	5.0	8.0	3.0	9.0
8	18.9	21.7	2.8	7.84
9	6.2	10.5	4.3	18.49
10	13.1	17.9	4.8	23.04
Sum			$\sum d = 33.5$	$\sum d^2 = 130.29$

Step 2 Calculate $t = \dfrac{\sum\limits_{i=1}^{n} d_i}{\sqrt{\dfrac{n \bullet \sum\limits_{i=1}^{n} d_i^2 - \left(\sum\limits_{i=1}^{n} d_i\right)^2}{n-1}}} = \dfrac{33.5}{\sqrt{\dfrac{10 \times 130.29 - 33.5^2}{9}}}$

$= \dfrac{33.5}{4.48} = 7.48$

Step 3 Compare to the critical value of t for a 2-tailed test with 9 degrees of freedom at $\alpha = 0.05$; $t(0.05, 9) = 2.262$. Since our value of t exceeds the critical value we conclude that the null hypothesis that the mean difference is zero – i.e. no effect of surgical hysterectomy – is rejected. Hence premature menopause has resulted in a significant increase in plasma lipoprotein A.

6.4 Comparison with a non-paired t-test

It is instructive to compare the results obtained above with those that would be obtained if we treated the data as *independent* observations and carried out an unpaired t-test.

For an unpaired t-test on the above data, t can be calculated (see Chapter 4: *Student's t-test*) as:

$$t = \frac{|\bar{x}_1 - \bar{x}_2|}{\sqrt{\left[\frac{(n_1 - 1)s_1^2 + (n_2 - 1)s_2^2}{n_1 + n_2 - 2}\right] \bullet \left[\frac{1}{n_1} + \frac{1}{n_2}\right]}}$$

Degrees of freedom $df = n_1 + n_2 - 2 = 18$

The mean value and standard deviation for the before-surgery lipoprotein A concentrations are 10.15 and 4.487, respectively, and for the after-surgery concentrations are 13.50 and 4.364, respectively.

$$t = \frac{13.50 - 10.15}{\sqrt{\left[\frac{9 \times 4.487^2 + 9 \times 4.364^2}{18}\right] \bullet \left[\frac{1}{10} + \frac{1}{10}\right]}}$$

$$= \frac{3.35}{\sqrt{19.59 \bullet 0.2}} = \frac{3.35}{\sqrt{3.918}} = \frac{3.35}{1.98} = 1.69$$

The critical value of t for a 2-tailed test with 18 degrees of freedom at $\alpha = 0.05$; $t(0.05, 18) = 2.10$. Since our value of t is lower than the critical value we conclude that the null hypothesis that the mean difference is zero – i.e. no effect of the hysterectomy – is accepted. Hence we would conclude that premature menopause has produced no significant effect on plasma lipoprotein A.

Why does the unpaired t-test give a different result to the paired t-test? By pairing the data we are eliminating the variability in the basal pre-surgery plasma lipoprotein A concentrations of the different patients. This example illustrates that the paired t-test has greater **power** – i.e. ability to reject a false null hypothesis – than the unpaired t-test.

PractiStat tutorial 6.1: paired *t*-test

Step 1   Click the OPEN button on the toolbar or select *Open data file* from the *File* menu and navigate to the 'Samples' folder on the PractiStat CD-ROM. Open the file titled 'LPA Before'. On your computer screen, the data from this file should now be displayed and the name 'LPA Before' should appear on the samples list.

Step 2   Once again, select the OPEN button on the toolbar and open the file titled 'LPA After'. The data from this file should now be displayed and the name 'LPA After' should appear on the samples list.

Step 3   Click the PAIR-T button on the toolbar or select *Paired t-test* from the *Statistics* menu and check that the '2-tailed' option is selected in the 'Options' area at the top right of the PractiStat window. Multi-select 'LPA Before' and 'LPA After' from the samples list (Control-click on PCs or Command-click on Macs).

Step 4   The viewing area should now show the results of a 2-tailed paired t-test. The calculated *t*-value is 7.477 and $t_{critical}$ at 9 degrees of freedom at the 0.05 significance level is 2.262. The P-value is listed as 'P < 0.05'. Therefore we can conclude that premature menopause has resulted in a significant increase in plasma lipoprotein A.

You can compare the results obtained above with those that would be obtained if we treated the data as independent observations and carried out an unpaired *t*-test.

Step 5   Click the T-TEST button on the toolbar or select *Student's t-test* from the *Statistics* menu. The viewing area should now show the results of a 2-tailed unpaired *t*-test. The calculated *t*-value is 1.693 and $t_{critical}$ at 18 degrees of freedom at the 0.05 significance level is 2.101. The P-value is listed as 'NS (P > 0.05)' We would conclude that premature menopause produces no significant effect on plasma lipoprotein A.

6.5 Assumptions underlying the paired *t*-test

The theoretical assumptions of the *t*-test for independent samples also apply to the paired *t*-test (see Chapter 4: *The Student's t-test*) so that the paired differences are normally distributed. If these assumptions are not met, then the non-parametric counterpart, the *Wilcoxon Signed-Rank test*, should be used instead (see Chapter 7). The derivation of *t* for the paired *t*-test is explained in Box 6.1.

Box 6.1: Derivation of t for the paired t-test

$$t = \frac{\text{Mean difference}}{\text{Standard error of the difference}} = \frac{\dfrac{\sum\limits_{i=1}^{n} d_i}{n}}{\dfrac{\text{Standard deviation of difference}}{\sqrt{n}}}$$

$$= \frac{\dfrac{\sum\limits_{i=1}^{n} d_i}{n}}{\sqrt{\dfrac{\sum\limits_{i=1}^{n} d_i^2 - \dfrac{\left(\sum\limits_{i=1}^{n} d_i\right)^2}{n}}{n(n-1)}}} = \frac{\sum\limits_{i=1}^{n} d_i}{n \bullet \sqrt{\dfrac{\sum\limits_{i=1}^{n} d_i^2 - \dfrac{\left(\sum\limits_{i=1}^{n} d_i\right)^2}{n}}{n(n-1)}}}$$

$$= \frac{\sum\limits_{i=1}^{n} d_i}{n \bullet \sqrt{\dfrac{n \bullet \sum\limits_{i=1}^{n} d_i^2 - \left(\sum\limits_{i=1}^{n} d_i\right)^2}{n^2(n-1)}}} = \frac{\sum\limits_{i=1}^{n} d_i}{\sqrt{\dfrac{n \bullet \sum\limits_{i=1}^{n} d_i^2 - \left(\sum\limits_{i=1}^{n} d_i\right)^2}{n-1}}}$$

The Wilcoxon Signed-Rank test

A non-parametric test for paired samples

Chapter 7 examines how the non-parametric Wilcoxon Signed-Rank test can be used for ordinal or ranked data, or where at least one sample has outliers, to determine whether the median of the observed differences (after a treatment, for example) deviates enough from zero to conclude that the treatment, etc. has had a significant effect. We first examine the advantages of excluding within-group variation from our analysis, and then show how W can be calculated.

7.1 Description

The **Wilcoxon Signed-Rank test** for matched pairs is used to determine whether a given treatment had a significant effect on a population. This **non-parametric test** is based on ranks and is appropriate for ordinal or **ranked data**. The Wilcoxon Signed-Rank test may also be used if, when comparing two samples, at least one sample contains outliers (see Chapter 3: *Histograms and box plots*). The Wilcoxon test determines whether the **median** of the observed differences after treatment deviates enough from zero to conclude that the treatment had a significant effect.

7.2 Uses

Like the *t*-test for paired samples (Chapter 6) the Wilcoxon Signed-Rank test has the advantage that much of the within-group variations attributable to the initial individual differences between subjects is excluded from the analysis. Since the two groups of observations to be compared are obtained from the same sample of subjects tested twice (before and after a treatment), instead of analysing the raw data from each group separately we look only at the *differences* between the pre-treatment and post-treatment values for each subject. By subtracting the first value from the second for each subject and then analysing only these paired differences we exclude all the variation in our data that results from the differing initial values of individual subjects.

7.3 Formulae

$$z = \frac{|W| - 0.5}{s_w}$$

where W is the sum of the signed ranks of the differences:

$$W = \sum_{i=1}^{n} R_i$$

and s_w is the population standard deviation of W:

$$s_w = \sqrt{\frac{n(n+1)(2n+1)}{6}}$$

7.3.1 Calculation of W

1 Calculate the differences d_i between the paired observations – i.e. $d_i = x_i - y_i$

2 Rank the differences low to high ignoring the sign

3 If $d_i = 0$ the *i*th pair is omitted from the analysis

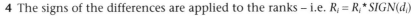

4 The signs of the differences are applied to the ranks – i.e. $R_i = R_i * SIGN(d_i)$

5 Sum the R_i values.

7.3.2 Correction for ties

When there are many ties in the ranks (ignoring sign), the standard deviation, s_w, needs to be reduced by an amount that is dependent on the number of ties:

$$s_w = \sqrt{\frac{n(n+1)(2n+1)}{6} - \sum_{i=1}^{n} \frac{(T_i-1)(T_i+1)}{12}}$$

where T_i is the number of ties in the ith rank

Note that when there are few ties, the reducing term in the correction is relatively small and has little effect on the calculated standard deviation s_w.

 Step-by-step tutorial 7.1: Wilcoxon Signed-Rank test

The plasma protein plasminogen-activator inhibitor type 1 (PAI-1) is an inhibitor of fibrinolysis in humans and may therefore be a factor in blood clot formation. Plasma levels of PAI increase in women after menopause, and this may contribute to the risk of cardiovascular disease. Investigators studied the effects of hormone-replacement therapy on PAI-1 levels on 10 post-menopausal women given oestrogen. The levels of PAI-1 (μg.l^{-1}) in the plasma before and after oestrogen treatment for each subject were as in Table 7.1.

Step 1 Calculate and tabulate the difference, rank of the difference (ignoring sign) and signed rank of the difference in value for each subject (Table 7.2).

Step 2 Calculate $W = \sum_{i=1}^{n} R_i = -10-6-1-4.5+4.5-9+2-3-7-8 = -42$

Step 3 Calculate $s_w = \sqrt{\frac{n(n+1)(2n+1)}{6}} = \sqrt{\frac{10(11)(21)}{6}} = \sqrt{385} = 19.6$

Note that since a single tie in the data has minimal effect on s_w we do not apply any correction for ties.

Step 4 Calculate

$$z = \frac{|W|-0.5}{s_w} = \frac{41.5}{19.6} = 2.12$$

Step 5 Compare this value of z with the critical value of z at the desired level of significance. Our calculated value is greater than the critical value of 1.96 for $P \leq 0.05$. Hence we can reject the null hypothesis and conclude that there is a significant effect of oestrogen on PAI-1 levels. Hormone replace-

Table 7.1 Levels of PAI-1 (μg.l^{-1}) in plasma

Subject	Before treatment	After treatment
1	80	50
2	19	12
3	23	22
4	54	50
5	8	12
6	59	30
7	12	14
8	6	3
9	40	32
10	34	17

Table 7.2 PAI levels: Calculating/tabulating difference and rank of difference

Subject	Before treatment	After treatment	Difference (d_i)	Rank of difference	Signed rank of difference (R_i)
1	80	50	−30	10	−10
2	19	12	−7	6	−6
3	23	22	−1	1	−1
4	54	50	−4	4.5	−4.5
5	8	12	+4	4.5	+4.5
6	59	30	−29	9	−9
7	12	14	+2	2	+2
8	6	3	−3	3	−3
9	40	32	−8	7	−7
10	34	17	−17	8	−8

$$W = \sum R_i = -42$$

ment therapy may therefore reduce the risk of cardiovascular disease in post-menopausal women.

 PractiStat tutorial 7.1: Wilcoxon Signed-Rank test

Step 1   Select the OPEN button on the toolbar or select *Open data file* from the *File* menu and navigate to the 'Samples' folder on the PractiStat CD-ROM. Open the file titled 'PAI Before'. On your computer screen, the data from this file should now be displayed and the name 'PAI Before' should appear on the samples list.

Step 2   Once again, select the OPEN button on the toolbar and open the file titled 'PAI After'. The data from this file should now be displayed and the name 'PAI After' should appear on the samples list.

Step 3   Select the WILCOX button on the toolbar or select *Wilcoxon Signed Rank Test* from the *Statistics* menu and check that the '2-tailed' option is selected in the 'Options' area at the upper right of the PractiStat window. Multi-select 'PAI Before' and 'PAI After' from the samples list (Control-click on PCs or Command-click on Macs).

Step 4   The viewing area should now show the results of a 2-tailed Wilcoxon Signed Rank test using the z-distribution. The calculated z-value is 2.115. The critical value of z in a 2-tailed test at the 0.05 significance level is 1.96. Since the calculated z-value exceeds $z_{critical}$, the P-value is listed as 'P < 0.05'. Therefore we can conclude that there was a significant effect of oestrogen on PAI-1 levels.

7.4 ▶ Assumption underlying the Wilcoxon Signed-Rank test

The paired values are randomly and independently drawn. Readers interested in the mathematical basis for the Wilcoxon Signed-Rank test should consult Box 7.1.

Box 7.1: More on the mathematics of the Wilcoxon Signed-Rank test

Since the sum of the first n integral numbers is $\dfrac{n(n+1)}{2}$ the maximum value that W can take – i.e. when all then n differences are positive – is $\dfrac{n(n+1)}{2}$ and the minimum value – i.e. when all the differences are negative – is $\dfrac{n(n+1)}{2}$. For any given value of $|W|$ between these extremes we can ask what proportion of the total number of ways of distributing the plus and minus signs between the n ranked differences would give rise to a value as large as the observed value. For the chance of such an occurrence to be ≤0.05 clearly there must be at least 20 arrangements. Since for any given value of n the number of *combinations* of plus and minus signs is 2^n, n must be at least 5 for it to be possible for a single combination to have only a 1 in 20 possibility of arising by chance. However for the purpose of illustration consider the case when n = 3. The various ways of arranging the plus and minus values together with the corresponding values of W are shown in the following table.

Continued

Box 7.1: *Continued*

Rank

1	+	−	+	+	−	−	+	−
2	+	+	−	+	−	+	−	−
3	+	+	+	−	+	−	−	−
W	6	4	2	0	0	−2	−4	−6

The possibility that a value of $W = 6$ has arisen by chance is therefore $1/8 = 0.125$. In general for n paired differences W will be distributed with a population mean of $\mu_W = 0$ and it can be shown that the population standard deviation:

$$s_W = \sqrt{\frac{n(n+1)(2n+1)}{6}}$$

As in the case of the Mann-Whitney U-test this value of s_W can be modified to take account of ties.

One-way Analysis of Variance (ANOVA)

Comparing the means of several groups

Chapter 8 considers how we proceed if we need to make comparisons between more than two sample means in a single test. We first explore how one-way Analysis of Variance (ANOVA) can be used to compare such means, and then show how between- and within-samples sum of squares, between- and within-samples degrees of freedom, total sum of squares and total degrees of freedom, between- and within-samples mean square and the F-ratio can be calculated. We finally show how to use F-tables to decide whether the samples are significantly different from each other.

8.1 Description

In Chapter 4 we saw how Student's t-test can be used to compare the mean values of 2 samples. But what happens when we have *several samples to compare*? It might be imagined that pairwise comparisons of the various samples with each other using multiple Student's t-tests would be a possible procedure. However, in addition to being a tedious computation, this is not desirable for sound statistical reasons. Suppose, for instance, that we have six samples to compare and we set our level of significance at 0.05. We would have to perform 15 pair-wise comparisons to arrive at 15 *simultaneous* justified conclusions. We know that for a single t-test the probability of rejecting a null hypothesis that is in fact true (this is known as a Type I error) is 1 in 20, but because our 15 conclusions are simultaneous, the overall risk of making a Type I error is much higher than 0.05 since we are concluding '*a*' AND '*b*' AND '*c*' AND . . . so on; in fact the chance that at least one Type I error will be made in the 15 t-tests can be shown to be 0.37. In order to compensate for this we could set the significance level at the more stringent value of 0.001, but then we would increase the risk of a Type II error – i.e. failing to reject a null hypothesis that is in fact untrue. Analysis of variance, usually referred to as ANOVA, overcomes this problem by allowing us to make comparisons between any number of sample means in a single test.

8.2 Uses

When ANOVA is used to compare the means of several samples it is referred to as **one-way ANOVA**. However more complicated analyses are also possible. **Two-way ANOVA**, for example, allows analysis of the effects of two independent variables on a dependent variable and is considered in Chapter 13. The analysis of matched samples by ANOVA is considered in Chapter 9.

Despite its name, **Analysis of Variance** is a method for comparing *means*. It tests the global null hypothesis, H_o, that for k population means, $\mu_1 \ldots \mu_k$:

$$H_o: \mu_1 = \mu_2 = \mu_3 = \ldots = \mu_k$$

If the null hypothesis is rejected it means that *at least* two of the population means are significantly different from each other. To decide exactly which ones are significantly different requires further analysis using the post hoc multiple comparison methods presented in Chapter 10: *Multiple comparison tests*.

8.3 Formulae

Let x_{ij} be the jth observation in the ith sample; let there be k samples and let n_i be the number of observations in the ith sample.

8.3.1   Between-samples sum of squares (SS$_B$)

$$SS_B = \sum_{i=1}^{k} \left(\frac{\sum\limits_{j=1}^{n_i} x_{ij}^2}{n_i} \right) - \frac{\left(\sum\limits_{i=1}^{k} \sum\limits_{j=1}^{n_i} x_{ij} \right)^2}{\sum\limits_{i=1}^{k} n_i}$$

If $\bar{x}_i$ is the mean value of the ith sample and $\bar{X}$ is the overall mean of all the observations the above equation can also be written as:

$$SS_B = \sum_{i=1}^{k} n_i \cdot \left(\bar{x}_i - \bar{X} \right)^2$$

8.3.2   Within-samples sum of squares (SS$_W$)

$$SS_W = \sum_{i=1}^{k} \sum_{j=1}^{n_i} x_{ij}^2 - \sum_{i=1}^{k} \left(\frac{\left(\sum\limits_{j=1}^{n_i} x_{ij} \right)^2}{n_i} \right)$$

The above equation can also be written as:

$$SS_W = \sum_{i=1}^{k} \sum_{j=1}^{n_i} \left(x_{ij} - \bar{x}_i \right)^2$$

8.3.3   Between-samples degrees of freedom (df$_B$)

$$df_B = k - 1$$

8.3.4   Within-samples degrees of freedom (df$_W$)

$$df_W = \sum_{i=1}^{k} n_i - k$$

8.3.5   Total sum of squares (SS$_T$)

$$SS_T = \sum_{i=1}^{k} \sum_{j=1}^{n} \left(x_{ij} - \bar{X} \right)$$

$$SS_T = SS_B + SS_W$$

8.3.6 Total degrees of freedom (df$_T$)

$$df_T = \sum_{i=1}^{k} n_i - 1$$

8.3.7 Between-samples mean square (MS$_B$)

$$MS_B = \frac{SS_B}{df_B}$$

8.3.8 Within-samples mean square (MS$_W$)

$$MS_W = \frac{SS_W}{df_W}$$

8.3.9 F-ratio (F)

$$F = \frac{MS_B}{MS_W}$$

8.4 Using the F-tables (Appendix 4)

From the F-tables in Appendix 4, locate the table for the set significance level (α). Next, locate the row corresponding to the within-samples degrees of freedom (df_W) then select the critical value of F at the column corresponding to the between-samples degrees of freedom (df_B). If the calculated value of F exceeds the critical value of F we can conclude that there is significant difference between at least two of the mean values. To decide which of the samples are significantly different from each other, use the most appropriate of the multiple comparison tests described in Chapter 10.

8.5 Theory of ANOVA

Fundamental to ANOVA is the idea that the null hypothesis can be rejected if the variability *between* the mean values of our samples is greater than can be accounted for by the intrinsic variability of the data *within* our samples. We therefore calculate quantities that express the variability of the data between the samples and variability of the data within the samples. For both kinds of variability we calculate the variability as a mean square deviation, of the general form:

$$\frac{\text{Sum of squares}}{\text{Degrees of freedom}} \quad \text{or} \quad \frac{\sum(x-\bar{x})^2}{n-1}$$

Under the null hypothesis, if the samples are drawn from normally distributed populations with equal means and variances, then the variance estimate based on within-group variability should be about the same as the variance due to between-groups variability. We compare these 2 estimates of variance via the F-test, which tests whether the ratio of the 2 estimates is close enough to 1 to conclude that the null hypothesis is true at the α level of significance:

$$F = \frac{\text{Between-samples variability}}{\text{Within-samples variability}}$$

 Step-by-step tutorial 8.1: detailed calculations for one-way ANOVA

We illustrate the calculations with a simple example. Suppose we measure the weight gain (in grams) of male laboratory rats on three different diets over a 4-week period. The results are as in Table 8.1.

Table 8.1 Weight gain of male laboratory rats

	Diet 1	Diet 2	Diet 3
	90	120	125
	95	125	130
	100	130	135
Mean weight gain (g):	95	125	130

Step 1 Calculate the within-samples sum of squares for each diet

Within-samples sum of squares:

$$SS = \sum(x-\bar{x})^2$$
$$SS_{\text{Diet 1}} = (90-95)^2 + (95-95)^2 + (100-95)^2 = 50$$
$$SS_{\text{Diet 2}} = (120-125)^2 + (125-125)^2 + (130-125)^2 = 50$$
$$SS_{\text{Diet 3}} = (90-95)^2 + (95-95)^2 + (100-95)^2 = 50$$

Total within–samples sum of squares: $50 + 50 + 50 = 150$

Step 2 Calculate the overall mean

Overall mean: $(90 + 95 + 100 + 120 + 125 + 130 + 125 + 130 + 135)/9 = 117$

Step 3 Calculate the total sum of squares

Total sum of squares: $(90 - 117)^2 + (95 - 117)^2 + (100 - 117)^2 + (120 - 117)^2 + (125 - 117)^2 + (130 - 117)^2 + (125 - 117)^2 + (130 - 117)^2 + (135 - 117)^2 = 2300$

Step 4 Calculate the between-samples sum of squares. Notice that the value for the between samples sum of squares can be arrived at in two different ways.

First we can use the formula: between-samples $SS = \sum_{i=1}^{k} n_i(\bar{x}_i - \bar{X})^2$

Where n_i is the number of observations in the ith sample, k is the number of samples and $\bar{X}$ is the overall mean:

Between-samples SS $\quad 3(95-117)^2 + 3(125-117)^2 + 3(130-117)^2 = 2150$

Alternatively we can calculate the between-samples SS as the difference between the total SS and the total within-samples SS

Between-samples SS $\qquad\qquad 2300 - 150 = 2150$

Step 5 Calculate degrees of freedom and mean squares

Within-samples degrees of freedom	$9 - 3 = 6$
Within-samples mean square	$150/6 = 25$

Between-samples degrees of freedom	2
Between-samples mean square	$2150/2 = 1075$

Step 6 Calculate the ratio:

$$F = \frac{\text{between-samples mean square}}{\text{within-samples mean square}}$$

and compare the value to the critical value of F from the tabulated values in Appendix 4 for $df_W = 6$, $df_B = 2$ and $\alpha = 0.05$. For the example given above $F = \dfrac{1075}{2} = 43$. The critical values of $F(df_W = 6; df_B = 2; \alpha = 0.05)$ is 5.14. Since the calculated value of 43 exceeds the critical value of 5.14 we can conclude that there is a significant effect of diet on the weight gain of male laboratory rats.

In Chapter 10 we shall show how to analyse which of the groups are significantly different from each other.

Step-by-step tutorial 8.2: one-way ANOVA

Leptin is a hormone secreted by fat cells which plays a role in regulation of body weight. A commonly used indicator of body weight is the Body Mass Index (BMI) which is calculated as one's weight in kilograms divided by the square of one's height in metres. A researcher measured blood leptin levels (ng/ml) in three groups of subjects: Group 1 (obese, body mass index (BMI) > 30 kg/m²), Group 2 (normal weight, BMI 25) and Group 3 (underweight; BMI < 20).

The data obtained are summarised in Table 8.2.

Step 1 We first calculate the mean leptin concentrations and within- and between-samples sums of squares for the three groups (Table 8.3).

Table 8.2 Blood leptin concentrations (ng/ml)

Group 1	Group 2	Group 3
34	38	12
15	20	5
23	15	32
50	19	19
40	22	25
22	34	6
32	25	25
34	35	34
41	28	19
36		31

Table 8.3 Mean leptin concentrations and within-/between-samples sum of squares

Data	Group 1	Group 2	Group 3	Total
Mean leptin (ng/ml)	32.7	26.2	20.8	
Within-samples SS	958	516	992	2466
Between-samples SS	374	1	335	710

Step 2 The within-samples and between-samples mean squares can then be calculated:

$$MS_w = \frac{SS_w}{df_W} = \frac{2466}{29-3} = 94.8$$

$$MS_B = \frac{SS_B}{df_B} = \frac{710}{3-1} = 355$$

$$\text{Hence } F = \frac{MS_B}{MS_W} = \frac{355}{94.8} = 3.74$$

Step 3 Since $F(2,26; 0.05)$ is 3.37 we can reject the null hypothesis and conclude that there exists a significant difference between the leptin values in the three groups. In Chapter 10 we shall show how to analyse which of the groups are significantly different from each other.

PractiStat tutorial 8.1: one-way ANOVA

Step 1 Select the OPEN button on the toolbar or select *Open data file* from the *File* menu and navigate to the 'Samples' folder on the PractiStat CD-ROM. Open the data file 'BMI Less-Than-20'. The data from this file should now be displayed and the name 'BMI Less-Than-20' should appear on the samples list.

Step 2 Similarly open the data files 'BMI Normal' and 'BMI Over-30'.

Step 3 Make sure all 3 data sets are selected (Control-click on PCs or Command-click on Macs).

Step 4 Select *Analysis of Variance* from the *Statistics* menu or click on the ANOVA button. The program should now show the ANOVA results for these samples. The calculated F-value is 3.743. The critical value of F at 2 degrees of freedom (between) and 26 degrees of freedom (within) and at the 0.05 significance level is 3.369. Since the calculated F-value exceeds $F_{critical}$, the P-value is listed as 'P < 0.05'. Therefore we can conclude that there was a significant difference between the leptin values in the three groups.

8.6 Assumptions of one-way ANOVA

1 The dependent variable is measured on an equal interval scale

2 The k samples are drawn independently and randomly from the source population(s)

3 The source population(s) are normally distributed

4 The k samples have approximately equal variances (ratio of largest to smallest variance not more than 1.5).

In practice it is found that the one-way ANOVA is fairly resistant to violation of assumptions 2 and 3 above, provided that the number of values in each sample is the same. It is therefore recommended that you try to ensure equal sample size when carrying out one-way ANOVA. If assumption 4 is not satisfied and the variances are very different then the non-parametric Kruskal-Wallis test described in Chapter 11 should be used instead of one-way ANOVA.

Matched-samples ANOVA

Comparing several matched groups

Chapter 9 examines how matched-samples ANOVA can be used to measure random variability in our data where our experimental design involves each subject being measured under three or more conditions. We show how within- and between-samples sum of squares, within-samples sum of squares corrected for pre-existing variability, degrees of freedom, between-samples and corrected within-samples mean square and the F-ratio can be calculated.

9.1 ▸ Description

The matched-samples ANOVA is similar to the paired t-test (Chapter 6) except that now the number of conditions is three or more. The matched-samples ANOVA is highly effective in removing the extraneous variations arising from pre-existing individual differences. The portion of the within-samples **sum of squares**, SS_W, that is attributable to pre-existing individual differences, designated as $SS_{subjects}$, is dropped from the analysis; and the portion that remains is then used as the measure of random variability.

9.2 ▸ Uses

In the paired t-test we either have a certain number of subjects, each measured under two conditions, A and B, or, alternatively, we have a certain number of matched pairs of subjects, and carry out measurements on one member of the pair under condition A and the other under condition B. The experimental design for matched-samples ANOVA is similar, except that now the number of conditions is 3 or more: $A|B|C$, $A|B|C|D$, etc. When the analysis involves each subject being measured under each of the k conditions, it is referred to as a **repeated measures** or **within-subjects design**. When the subjects are matched in sets of 3 for $k = 3$, 4 for $k = 4$, etc. and the subjects in each matched set are randomly assigned to one or another of the k conditions, it is described as a **randomised blocks design**, each set of k matched subjects being called a 'block'.

9.3 ▸ Formulae

If there are n subjects each tested under k conditions, x_{ij} is the jth value for the ith subject, $\bar{x}_i$ is the mean value of the ith subject and $\bar{X}$ is the overall mean value then:

9.3.1 Subjects sum of squares (SS$_{subjects}$)

$$SS_{subjects} = k \sum_{i=1}^{n} \left(\bar{x}_i - \bar{X} \right)^2$$

9.3.2 Total sum of squares (SS$_T$)

$$SS_T = \sum_{i=1}^{k} \sum_{j=1}^{n_i} \left(\bar{x}_{ij} - \bar{X} \right)^2$$

9.3.3 Within-samples sum of squares (SS_W)

$$SS_W = \sum_{i=1}^{k} \sum_{j=1}^{n_i} (x_{ij} - \bar{x}_i)^2$$

9.3.4 Between-samples sum of squares (SS_B)

$$SS_B = SS_T - SS_W$$

9.3.5 Within-samples sum of squares corrected for pre-existing variability (SS_W^corr)

$$SS_{Wcorr} = SS_W - SS_{subjects}$$

9.3.6 Degrees of freedom

$$df_{total} = nk - 1$$
$$df_W = nk - k$$
$$df_B = k - 1$$
$$df_{subjects} = n - 1$$
$$df_{Wcorr} = df_W - df_{subjects}$$

9.3.7 Between-samples mean square (MS_B)

$$MS_B = \frac{SS_B}{df_B}$$

9.3.8 Corrected within-samples mean square

$$MS_{Wcorr} = \frac{SS_{Wcorr}}{df_{Wcorr}}$$

9.3.9 F-ratio (F)

$$F = \frac{MS_B}{MS_{Wcorr}}$$

Table 9.1 Immune response to an antigen

	Response to antigen preparation (arbitrary units)			
	1	2	3	Subject means
	12	16	24	17.33
	50	60	100	70.00
	4	13	20	12.33
	34	45	50	43.00
	100	150	200	150
	40	38	37	38.33
	10	10	12	10.67
	20	60	50	43.33
	23	56	20	33.00
	90	120	100	103.33
Group means	38.30	56.80	61.30	**Overall mean** = 52.13

Step-by-step tutorial 9.1: matched-samples ANOVA

An immunologist was studying the immune response to a particular antigen. Three different preparations of antigen were each administered to 10 subjects on three separate occasions and the antibody titre (arbitrary units) measured. The results are given in Table 9.1, which also shows the group means and the subject means.

Step 1 Calculate the sums of squares $\Sigma(x - \bar{x})^2$ for the 3 different groups:

$$SS_1 = (12 - 38.3)^2 + (50 - 38.3)^2 + (4 - 38.3)^2 + (34 - 38.3)^2 + (100 - 38.3)^2$$
$$+ (40 - 38.3)^2 + (10 - 38.3)^2 + (20 - 38.3)^2 + (23 - 38.3)^2 + (90 - 38.3)^2$$
$$= 9876$$

$$SS_2 = (16 - 56.8)^2 + (60 - 56.8)^2 + (13 - 56.8)^2 + (45 - 56.8)^2 + (150 - 56.8)^2$$
$$+ (38 - 56.8)^2 + (10 - 56.8)^2 + (60 - 56.8)^2 + (56 - 56.8)^2 + (120 - 56.8)^2$$
$$= 18967$$

$$SS_3 = (24 - 61.3)^2 + (100 - 61.3)^2 + (20 - 61.3)^2 + (50 - 61.3)^2 + (200 - 61.3)^2$$
$$+ (37 - 61.3)^2 + (12 - 61.3)^2 + (50 - 61.3)^2 + (20 - 61.3)^2 + (100 - 61.3)^2$$
$$= 30312$$

Step 2 Calculate the total sum of squares:

$$SS_T = (12 - 52.13)^2 + (16 - 52.13)^2 + (24 - 52.13)^2 + (50 - 52.13)^2$$
$$+ (60 - 52.13)^2 + (100 - 52.13)^2 + (4 - 52.13)^2 + (13 - 52.13)^2$$
$$+ (20 - 52.13)^2 + (34 - 52.13)^2 + (45 - 52.13)^2 + (50 - 52.13)^2$$
$$+ (100 - 52.13)^2 + (150 - 52.13)^2 + (200 - 52.13)^2 + (40 - 52.13)^2$$

$$+(38-52.13)^2+(37-52.13)^2+(10-52.13)^2+(10-52.13)^2$$
$$+(12-52.13)^2+(20-52.13)^2+(60-52.13)^2+(50-52.13)^2$$
$$+(23-52.13)^2+(56-52.13)^2+(20-52.13)^2+(90-52.13)^2$$
$$+(120-52.13)^2+(100-52.13)^2$$
$$=62127$$

Step 3 Calculate the within-samples sum of squares:

$$SS_W = SS_1 + SS_2 + SS_3 = 59155$$

Step 4 Calculate the between-samples sum of squares:

$$SS_B = SS_T - SS_W = 2972$$

Notice that SS_B is small in comparison with SS_W and if you were performing the analysis as an independent-samples ANOVA, this would give a non-significant result, since the large value of SS_W would also give you a large value of MS_W and hence a low value of F. However because the samples are matched we can eliminate the variability arising from pre-existing differences in the subjects as follows.

Step 5 Calculate the subjects sum of squares:

$$SS_{subjects} = k\sum_{i=1}^{n}(\bar{x}_i - \bar{X})^2$$

$$SS_{subjects} = 3 \times \big((17.33-52.13)^2+(70.00-52.13)^2+(12.33-52.13)^2$$
$$+(43.00-52.13)^2+(150.00-52.13)^2+(38.33-52.13)^2$$
$$+(10.67-52.13)^2+(43.33-52.13)^2+(33.00-52.13)^2$$
$$+(103.33-52.13)^2\big)$$
$$=53252$$

Step 6 Calculate the within-samples sum of squares corrected for pre-existing variability:

$$SS_{W^{corr}} = SS_W - SS_{subjects} = 59155 - 53252 = 5903$$

Step 7 Calculate the degrees of freedom:

$$df_B = k - 1 = 3 - 1 = 2$$
$$df_{subjects} = n - 1 = 10 - 1 = 9$$
$$df_W = nk - k = 30 - 3 = 27$$
$$df_{W^{corr}} = df_W - df_{subjects} = 27 - 9 = 18$$

Step 8 Calculate the mean square values:

$$MS_B = \frac{SS_B}{df_B} = \frac{2972}{2} = 1486$$

$$MS_{Wcorr} = \frac{SS_{Wcorr}}{df_{Wcorr}} = \frac{5903}{18} = 328$$

Step 9   Calculate $F = \dfrac{MS_B}{MS_{Wcorr}} = \dfrac{1486}{328} = 4.53$

Since $F(2,18; 0.05)$ is 3.55 (Appendix 4) we can reject the null hypothesis and conclude that there exists a significant difference between the values in the three groups.

 PractiStat tutorial 9.1: matched-samples ANOVA

Step 1   Select the OPEN button on the toolbar or select *Open data file* from the *File* menu and navigate to the 'Samples' folder on the PractiStat CD-ROM. Open the data file 'Antigen Prep-1'. The data from this file should now be displayed and the name 'Antigen Prep-1' should appear on the samples list.

Step 2   Similarly open the data files 'Antigen Prep-2' and 'Antigen Prep-3'.

Step 3   Make sure all three data sets are selected (Control-click on PCs or Command-click on Macs).

Step 4   Select *Repeated Measures ANOVA* from the *Statistics* menu or click on the ANOVA-R button on the toolbar. The program should now show the Repeated Measures ANOVA results for these samples. The calculated *F*-value is 4.530. The critical value of *F* at 2 degrees of freedom (between) and 18 degrees of freedom (within) and at the 0.05 significance level is 3.555. Since the calculated *F*-value exceeds $F_{critical}$, the P-value is listed as 'P < 0.05'. Therefore we can conclude that there exists a significant difference between the values in the three groups.

 # 9.4 Assumptions of matched-samples ANOVA

Assumptions 1–4 are similar to those for the independent-samples ANOVA:

1 The dependent variable is measured on an equal interval scale

2 The measures within each of the *k* groups are independent of each other

3 The source population(s) from which the *k* samples are drawn are normally distributed

4 The *k* groups have approximately equal variances

5 All possible correlation coefficients (*r*) among the *k* groups of measures are positive and of similar magnitude.

We noted in Chapter 8 that the analysis of variance is quite resistant to violations of assumptions 3 and 4, providing that the *k* groups have the same size. In the matched-samples ANOVA this condition is clearly always satisfied. Condition 5 is unique to the matched-samples version of ANOVA. The calculation of correlation coefficients is described in Chapter 14.

10

Multiple comparison tests

Post hoc methods for comparing the means of several groups

Chapter 10 gives an overview of post hoc testing methods and the test statistics for pair-wise comparisons of all the samples in an ANOVA. Although a significant F-value from one-way ANOVA can tell us that one mean is significantly different from another mean in the analysis, we need multiple comparisions procedures to find out which means are significantly different from each other. In a priori tests comparisons are made on a subset of the data decided on before the experiment. We describe here four of the more common post hoc tests where the decision on which groups to test is made after the data are collected – Bonferroni's t-test, Fisher's LSD test, the Tukey test and the Control t-test.

10.1 Overview of post hoc testing

A significant F-value from one-way ANOVA tells you that at least one mean is significantly different from another mean in the analysis but does not indicate which means are significantly different from each other. The procedures used to examine these differences are called **Multiple comparison tests**. When comparisons are made only on a subset of the data *as decided before the experiment* the tests are referred to as **a priori tests**. More commonly decisions as to which groups to test are made only after the data are collected and take account of *all* the data sets. These tests are referred to as **post hoc tests**, and in this chapter we describe the use of 4 common post hoc tests. It should be noted that there are many more post hoc tests available that are not covered here and there is no universal agreement as to which test is best.

The main problem that post hoc tests address is that the more tests that you conduct at, say, $\alpha = 0.05$, the more likely you are to make a Type I error – i.e. conclude erroneously that a significant difference exists. If, for example, you carry out all the 10 possible t-tests on 5 means at $\alpha = 0.05$ the overall chance of making a Type I error is $1 - (1 - 0.05)^{10} = 0.401$. In other words there is a 40% chance of a Type I error instead of the desired 5%.

10.2 Post hoc testing methods

Bonferroni's t-test and the **Control t-test** take account of the problem of multiple tests by applying **Bonferroni's correction** to the calculated probability – i.e. the critical value of t for each comparison is determined at a P-value that is the desired P-value divided by a factor dependent on the number of tests possible. Fisher's LSD (Least Significant Difference) test adjusts the t-statistic by a factor related to the within samples sum of squares and the size of the samples. The Tukey test uses the Tukey distribution instead of Student's t. Both Bonferroni's, Fisher's LSD and the Tukey tests are appropriate for testing all possible pairs of values; the Control t-test is used when you wish to test one control value against all the other samples.

10.3 Bonferroni's t-test

10.3.1 Description

Bonferroni's t-test is a post hoc analysis that does pair-wise comparisons of *all* the samples in an ANOVA. For each pair of samples, the Student's t-test is employed but with the **Bonferroni inequality** applied to the calculated probability.

Within the context of multiple comparisons, the Bonferroni inequality states that if k pair-wise comparisons are made, then the probability of erroneously concluding that any two samples are different from each other increases k-fold.

For example, if in a 4-sample analysis (i.e. 6 pair-wise comparisons) the Student's t-test for A vs. B yields P = 0.01, then Bonferroni's t-test will yield P = $k \times$ P = $6 \times 0.01 = 0.06$.

Post hoc tests should be applied only after the ANOVA has yielded a significant difference.

10.3.2 Uses

When the results of an ANOVA analysis indicate that significant difference(s) exist, the Bonferroni t-test can be used to test the significance of the differences between the mean values of the various groups.

10.3.3 Formulae

Number of comparisons (k)

$$k = \frac{m(m-1)}{2}$$

where $m =$ is the total number of groups

Student's t-test

$$t = \frac{|\bar{x}_1 - \bar{x}_2|}{\sqrt{\left[\frac{(n_1 - 1)s_1^2 + (n_2 - 1)s_2^2}{n_1 + n_2 - 2}\right] \bullet \left[\frac{1}{n_2} + \frac{1}{n_2}\right]}}$$

Degrees of freedom

Degrees of freedom: $df = n_1 + n_2 - 2$

Bonferroni's inequality on the P-value

$$P_B = \frac{P}{k}$$

Step-by-step tutorial 10.1: Bonferroni's *t*-test

In Chapter 8 we carried out ANOVA analysis on the results of measurements of blood leptin levels in 3 groups of patients. The analysis indicated that there existed significant difference(s) between the mean values of the 3 groups. We shall now use Bonferroni's *t*-test to determine *which* groups are significantly different.

For convenience we give again the leptin concentrations (ng/ml) for the 3 groups (Table 10.1).

Table 10.1 Blood leptin concentrations for analysis by Bonferroni's t-test

Group 1	Group 2	Group 3
34	38	12
15	20	5
23	15	32
50	19	19
40	22	25
22	34	6
32	25	25
34	35	34
41	28	19
36		31

Step 1 Calculate the means and standard deviations of the 3 samples (see Chapter 2: *Descriptive statistics*) (Table 10.2).

Table 10.2 Leptin concentrations: calculating means and standard deviations

Group	Mean	SD	*n*
1	32.7	10.318	10
2	26.222	8.028	9
3	20.8	10.497	10

Step 2 Calculate *t*-values (see Chapter 4: *The Student's t-test*) for all possible comparisons.

Group 1 vs Group 2

$$t = \frac{32.7 - 26.222}{\sqrt{\left[\frac{9 \times 10.318^2 + 8 \times 8.028^2}{10 + 9 - 2}\right] \cdot \left[\frac{1}{10} + \frac{1}{9}\right]}} = 1.514$$

To test whether this value is significant we apply Bonferroni's inequality. For P < 0.05 we need to exceed the critical *t*-value for P/k = 0.05/3 = 0.0167. Although the *t*-table does not give values for P < 0.0167 for a 2-tailed test we note that $t(17; 0.02) = 2.567$. Since our value of *t* is less than this critical value the difference between Groups 1 and 2 is not significant at the 5% level.

Group 1 vs Group 3

$$t = \frac{32.7 - 20.8}{\sqrt{\left[\frac{9 \times 10.318^2 + 9 \times 10.497^2}{10 + 10 - 2}\right] \cdot \left[\frac{1}{10} + \frac{1}{10}\right]}} = 2.557$$

To test whether this value is significant we apply Bonferroni's inequality. For P < 0.05 we need to exceed the critical *t*-value for P/k = 0.05/3 = 0.0167. Although the *t*-table does not give values for P < 0.0167 for a 2-tailed test we note that $t(18; 0.02) = 2.552$. Hence the difference between Groups 1 and 3 is very close to being significant at the 5% level.

Group 2 vs Group 3

$$t = \frac{26.222 - 20.8}{\sqrt{\left[\frac{8 \times 8.028^2 + 9 \times 10.497^2}{9 + 10 - 2}\right] \cdot \left[\frac{1}{9} + \frac{1}{10}\right]}} = 1.253$$

To test whether this value is significant we apply Bonferroni's inequality. For P < 0.05 we need to exceed the critical *t*-value for P/k = 0.05/3 = 0.0167. Although the *t*-table does not give values for P < 0.0167 for a 2-tailed test we note that $t(17; 0.02) = 2.567$. Since our value of *t* is less than this critical value the difference between Groups 1 and 2 is not significant at the 5% level.

Note that although the one-way ANOVA results showed a difference in the means at the 0.05 significance level, the Bonferroni results show that none of the 3 pair-wise comparisons could be said to be different at the 0.05 significance level. This apparent contradiction is a result of the superior **power** (see Chapter 1: *Basic concepts of statistics*) of the Analysis of Variance to detect differences where they exist.

 PractiStat tutorial 10.1: Bonferroni's *t*-test

Step 1   Select the OPEN button on the toolbar or select *Open data file* from the *File* menu and navigate to the 'Samples' folder on the PractiStat CD-ROM. Open the data file 'BMI Less-Than-20'. The data from this file should now be displayed and the name 'BMI Less-Than-20' should appear on the samples list.

Step 2   Similarly open the data files 'BMI Normal' and 'BMI Over-30' from PractiStat's 'Samples' folder.

Step 3   Make sure all three data sets are selected in the samples list (Control-click on PCs or Command-click on Macs).

Step 4   Select *Analysis of variance* from the *Statistics* menu or click on the ANOVA button on the toolbar. The analysis results should now show 'P < 0.05'. There would be no point in continuing on to Bonferroni's Post Hoc if the ANOVA result was not significant.

Step 5   Select *Bonferroni's t-tests* from the *Statistics* menu or click on the BONF-T button on the toolbar. The program should now show all the possible pair-wise comparisons and their P-values.

Step 6   Note that although the one-way ANOVA results showed a difference in the means at the 0.05 significance level, the Bonferroni results show that none of the 3 pair-wise comparisons could be said to be different at the 0.05 significance level. As mentioned above, this apparent contradiction is a result of the superior **power** (see Chapter 1: *Basic concepts of statistics*) of the Analysis of Variance to detect differences where they exist. In the following section, we will see how Fisher's LSD test, a post hoc test with more power than Bonferroni's test, shows where the significant difference in the group means occurred.

◆10.4◆ Fisher's LSD test

10.4.1   Description

The **Fisher Least Significant Difference (LSD)** test is a post hoc analysis that does pair-wise comparisons of all the samples in an ANOVA. This test should be used only after the Analysis of Variance (ANOVA) has yielded a significant difference. The test calculates a modified *t*-statistic based on the within samples mean square.

10.4.2   Uses

When the results of an ANOVA analysis indicate that significant difference(s) exist, Fisher's LSD test, also known as the Protected *t*-test, can be used to test the significance of the differences between the mean values of the various groups.

10.4.3 Formulae

Number of comparisons, k, for m samples

$$k = \frac{m(m-1)}{2}$$

t-test

$$t = \frac{|\bar{x}_1 - \bar{x}_2|}{\sqrt{MS_w\left[\dfrac{1}{n_1} + \dfrac{1}{n_2}\right]}}$$

where MS_w is the within-samples mean square (see Chapter 8) and $\bar{x}_1$ and $\bar{x}_2$ are the mean values for two groups with n_1 and n_2 values respectively.

Degrees of freedom, df

$$df = \sum_{i=1}^{m} n_i - m$$

Where $\displaystyle\sum_{i=1}^{m} n_i$ is the total number of observations

Step-by-step tutorial 10.2: Fisher's LSD test

We illustrate the procedure using the data on leptin concentrations in 3 groups of subjects that we analysed in Chapter 8 by one-way ANOVA.

Step 1 The results of the ANOVA as calculated in Chapter 8 are given in Table 10.3.

Step 2 The within-samples mean square is calculated as:

$$MS_w = \frac{SS_w}{df_w} = \frac{2466}{29-3} = 94.8$$

Step 3 There are $\dfrac{3 \times 2}{2} = 3$ pair-wise comparisons to make: 1 vs 2; 1 vs 3; 2 vs 3.

Table 10.3 ANOVA results for blood leptin concentrations for analysis by Fisher LSD test

Data	Group 1	Group 2	Group 3	Total
n	10	9	10	29
Mean leptin (ng/ml)	32.7	26.2	20.8	
Within-samples SS	958	516	992	2466
Between-samples SS	374	1	335	710

Step 4   We calculate the *t*-statistic $t = \dfrac{|\overline{X}_1 - \overline{X}_2|}{\sqrt{MS_w\left[\dfrac{1}{n_1} + \dfrac{1}{n_2}\right]}}$ for each of these in turn.

Group 1 vs Group 2

$$t = \frac{|32.7 - 26.2|}{\sqrt{94.8 \bullet \left[\dfrac{1}{10} + \dfrac{1}{9}\right]}} = 1.453$$

Compare this value with $t(0.05; 26) = 1.706$. Since our calculated value is smaller than this critical value we conclude that there is no significant difference between the mean values of Groups 1 and 2.

Group 1 vs Group 3

$$t = \frac{|32.7 - 20.8|}{\sqrt{94.8 \bullet \left[\dfrac{1}{10} + \dfrac{1}{10}\right]}} = 2.733$$

This time our calculated *t*-statistic is greater than $t(0.05; 26)$ and we therefore conclude that Groups 1 and 3 show a statistically significant difference.

Group 2 vs Group 3

$$t = \frac{|26.2 - 20.8|}{\sqrt{94.8 \bullet \left[\dfrac{1}{9} + \dfrac{1}{10}\right]}} = 1.202$$

This time our calculated *t*-statistic is less than $t(0.05; 26)$ and we therefore conclude that the means of Groups 2 and 3 are not significantly different.

 PractiStat tutorial 10.2: Fisher's LSD test

Step 1   Select the OPEN button on the toolbar or select *Open data file* from the *File* menu and navigate to the 'Samples' folder on the PractiStat CD-ROM. Open the data file 'BMI Less-Than-20'. The data from this file should now be displayed and the name 'BMI Less-Than-20' should appear on the samples list.

Step 2   Similarly open the data files 'BMI Normal' and 'BMI Over-30' from PractiStat's 'Samples' folder.

Step 3   Make sure all 3 data sets are selected in the samples list (Control-click on PCs or Command-click on Macs).

Step 4   Select *Analysis of variance* from the *Statistics* menu or click on the ANOVA button on the toolbar. The analysis results should now show 'P < 0.05'. There would be no point in continuing on to the Fisher's LSD Post Hoc if the ANOVA result was not significant.

Step 5 Select *Fisher's LSD* from the *Statistics* menu or click on the FISHER button on the toolbar. The program should now show all the possible pair-wise comparisons and their P-values. The results show that 'BMI Less-Than-20' is significantly different from 'BMI Over-30' at the 0.05 level.

10.5 Tukey test

10.5.1 Description

The **Tukey test** is a post hoc analysis that does pair-wise comparisons of all the samples in an ANOVA. This test should be used only after the Analysis of Variance (ANOVA) has yielded a significant difference. The Tukey test compares the differences between the sample means with critical values determined from the Tukey distribution.

10.5.2 Uses

When the results of an ANOVA analysis indicate that significant difference(s) exist, the Tukey test can be used to test the significance of the differences between the mean values of the various groups.

10.5.3 Formulae

Absolute value of difference between means of samples *i* and *j*

$$|\bar{x}_i - \bar{x}_j|$$

Number of comparisons, *k*, for *m* samples

$$k = \frac{m(m-1)}{2}$$

Degrees of freedom (df_w)

$$df_w = \sum_{i=1}^{m} n_i - m$$

Critical difference (T_{ij})

$$T_{i,j} = q\sqrt{\frac{MS_w\left(\dfrac{1}{n_i} + \dfrac{1}{n_j}\right)}{2}}$$

where MS_w is the within-samples mean square (see Chapter 8), n_i is the number of observations in sample i, q is a value obtained from the Tukey table (Appendix 5) for m samples and df_w degrees of freedom.

 ## Step-by-step tutorial 10.3: Tukey test

We illustrate the procedure using the data on leptin concentrations in three groups of subjects that we analysed in Chapter 8 by one-way ANOVA. The results of the ANOVA as calculated in Chapter 8 are given in Table 10.4.

Table 10.4 ANOVA results for blood leptin concentrations for analysis by the Tukey test

Data	Group 1	Group 2	Group 3	Total
n	10	9	10	29
Mean leptin (ng/ml)	32.7	26.2	20.8	
Within-samples SS	958	516	992	2466
Between-samples SS	374	1	335	710

Step 1 The within-samples mean square is calculated as:

$$MS_w = \frac{SS_w}{df_w} = \frac{2466}{29-3} = 94.8$$

Step 2 The Tukey test table (Appendix 5) does not list values of q for 26 degrees of freedom but instead gives values for 24 and 30 degrees of freedom. We can estimate the values of q at 26 degrees of freedom by simple proportions (**interpolation**). From the table's column for comparing 3 means ($a = 3$), the values of q for 24 and 30 degrees of freedom (n) are 3.53 and 3.49, respectively. The value for $q(a = 3; n = 26)$ lies somewhere between 3.53 and 3.49 and can be calculated using simple proportions:

$$\frac{q-3.49}{3.53-3.49} = \frac{26-30}{24-30}$$

or:

$$q = \frac{26-30}{24-30} \bullet (3.53-3.49) + 3.49 = 3.51$$

Step 3 There are $\frac{3 \times 2}{2} = 3$ pair-wise comparisons to make: 1 vs 2; 1 vs 3; 2 vs 3. We calculate the absolute difference in means and critical value T_{ij} for each of these in turn.

Group 1 vs Group 2

$$|\bar{x}_1 - \bar{x}_2| = |32.7 - 26.2| = 6.5$$

$$T_{1,2} = 3.51\sqrt{\frac{94.8}{2}\left(\frac{1}{10} + \frac{1}{9}\right)} = 11.1$$

Since the difference between means is smaller than this critical value we conclude that there is no significant difference between the mean values of Groups 1 and 2.

Group 1 vs Group 3

$$|\bar{x}_1 - \bar{x}_3| = |32.7 - 20.8| = 11.9$$

$$T_{1,3} = 3.51\sqrt{\frac{94.8}{2}\left(\frac{1}{10} + \frac{1}{10}\right)} = 10.81$$

Since the difference between means is greater than this critical value we conclude that there is a significant difference between the mean values of Groups 1 and 3.

Group 2 vs Group 3

$$|\bar{x}_2 - \bar{x}_3| = |26.2 - 20.8| = 5.4$$

$$T_{2,3} = 3.51\sqrt{\frac{98.4}{2}\left(\frac{1}{9} + \frac{1}{10}\right)} = 11.1$$

Since the difference between means is smaller than this critical value we conclude that there is no significant difference between the mean values of Groups 2 and 3.

 PractiStat tutorial 10.3: Tukey test

Step 1 Select the OPEN button on the toolbar or select *Open data file* from the *File* menu and navigate to the 'Samples' folder on the PractiStat CD-ROM. Open the data file 'BMI Less-Than-20'. The data from this file should now be displayed and the name 'BMI Less-Than-20' should appear on the samples list.

Step 2 Similarly open the data files 'BMI Normal' and 'BMI Over-30' from PractiStat's 'Samples' folder.

Step 3 Make sure all 3 data sets are selected in the samples list (Control-click on PCs or Command-click on Macs).

Step 4 Select *Analysis of variance* from the *Statistics* menu or click on the ANOVA button on the toolbar. The analysis results should now show 'P < 0.05'. There would be no point in continuing on to the Tukey test if the ANOVA result was not significant.

Step 5 Select *Tukey Test* from the *Statistics* menu or click on the TUKEY button on the toolbar. The program should now show all the possible pair-wise comparisons and their P-values. The results show that 'BMI Less-Than-20' is significantly different from 'BMI Over-30' at the 0.05 level.

10.6 Control *t*-test

10.6.1 Description

The Control *t*-test is a post hoc analysis that does pair-wise comparisons between a control and the other samples in an ANOVA. For each comparison to the control sample, the Student's *t*-test is employed, but with **Bonferroni's inequality** applied to the calculated probability.

10.6.2 Uses

When the results of an ANOVA analysis indicate that significant difference(s) exist and one group can be designated the control group, the Control *t*-test can be used to test the significance of the differences between the mean value of the control group and the means of the other groups.

10.6.3 Formulae

Number of comparisons (*k*)

$$k = m - 1$$

Where m – is the total number of groups

Student's *t*-test

$$t = \frac{|\bar{x}_1 - \bar{x}_2|}{\sqrt{\frac{(n_1 - 1)s_1^{2} + (n_2 - 1)s_2^{2}}{n_1 + n_2 - 2} \cdot \left[\frac{1}{n_2} + \frac{1}{n_2} \right]}}$$

Degrees of freedom

Degrees of freedom: $df = n_1 + n_2 - 2$

Bonferroni's inequality on the P-value

$$P_B = \frac{P}{k}$$

Step-by-step tutorial 10.4: Control *t*-test

In Chapter 8 we carried out ANOVA analysis on the results of measurements of blood leptin levels in 3 groups of patients. The analysis indicated that there existed significant difference(s) between the mean values of the 3 groups. We shall now use the Control *t*-test to determine *which* groups are significantly different. We shall designate Group 1, the obese group, as the control group and test the null hypotheses that Group 1 does not differ in mean values from Groups 2 and 3.

For convenience we give again the leptin values (ng/ml) for the three groups (Table 10.5).

Table 10.5 Blood leptin concentrations for analysis by the Control *t*-test

Group 1	Group 2	Group 3
34	38	12
15	20	5
23	15	32
50	19	19
40	22	25
22	34	6
32	25	25
34	35	34
41	28	19
36		31

Step 1 Calculate the means and standard deviations of the three samples (see Chapter 2: *Descriptive statistics*) (Table 10.6).

Table 10.6 Leptin concentrations: calculating means and standard deviations

Group	Mean	SD	n
1	32.7	10.318	10
2	26.222	8.028	9
3	20.8	10.479	10

Step 2 Calculate *t*-values (see Chapter 4: *Student's T-test*).

Group 1 vs Group 2

$$t = \frac{32.7 - 26.222}{\sqrt{\left[\dfrac{9 \times 10.318^2 + 8 \times 8.028^2}{10 + 9 - 2}\right] \bullet \left[\dfrac{1}{10} + \dfrac{1}{9}\right]}} = 1.514$$

To test whether this value is significant we apply Bonferroni's inequality. For P < 0.05 we need to exceed the critical t-value for $P/k = 0.05/2 = 0.025$. Although the t-table does not give values for P < 0.025 for a 2-tailed test, we note that $t(18;0.02) = 2.552$. Since our value of t is less than this critical value we can conclude that there is no significant difference (P > 0.05) between Groups 1 and 2.

Group 1 vs Group 3

$$t = \frac{32.7 - 20.8}{\sqrt{\left[\dfrac{9 \times 10.318^2 + 9 \times 10.497^2}{10 + 10 - 2}\right] \bullet \left[\dfrac{1}{10} + \dfrac{1}{10}\right]}} = 2.557$$

To test whether this value is significant we apply Bonferroni's inequality. For P < 0.05 we need to exceed the critical t-value for $P/k = 0.05/2 = 0.025$. Although the t-table does not give values for P < 0.025 for a 2-tailed test, we note that $t(18;0.02) = 2.552$. Since our value of t exceeds this critical value we can conclude that there is a significant (P < 0.05) difference between Groups 1 and 3.

 PractiStat tutorial 10.4: Control t-test

Step 1 Select the OPEN button on the toolbar or select *Open data file* from the *File* menu and navigate to the 'Samples' folder on the PractiStat CD-ROM. Open the data file 'BMI Less-Than-20'. The data from this file should now be displayed and the name 'BMI Less-Than-20' should appear on the samples list.

Step 2 Similarly open the data files 'BMI Normal' and 'BMI Over-30' from PractiStat's 'Samples' folder.

Step 3 Make sure all 3 data sets are selected in the samples list (Control-click on PCs or Command-click on Macs).

Step 4 Select *Analysis of variance* from the *Statistics* menu or click on the ANOVA button on the toolbar. The analysis results should now show 'P < 0.05'. There would be no point in continuing on to the Control t-test Post Hoc if the ANOVA result was not significant.

Step 5 Select *Control t-tests* from the *Statistics* menu or click on the CONTROL button on the toolbar. The program should now show a blank results window with a '⟨no control marked⟩' message.

Step 6 Click on 'BMI Over-30' in the samples list and press the 'Mark as Control' button above the samples list. The small icon to the left of the marked sample will be changed to indicate that sample as the control sample. (To unmark a control sample, click on the sample and press the 'Mark as Control' button again.)

Step 7   With the 'BMI Over-30' sample marked as the control, multi-select 'BMI Norm' and 'BMI Less-Than-20'. The program will now show the two pair-wise comparisons against 'BMI Over-30' and their P-values. The results show that 'BMI Less-Than-20' is significantly different from 'BMI Over-30' at the 0.05 level.

11

The Kruskal–Wallis test

A non-parametric method for comparing the means of several groups

Chapter 11 examines how the non-parametric Kruskal–Wallis test can allow us to make comparisons between any number of sample medians in a single text. We first show how interpretation of the Kruskal–Wallis test is based on ranks rather than means, and then how the H-statistic is calculated.

11.1 ▶ Description

In Chapter 5 we saw how the Mann–Whitney U-test can be used to compare the median values of two samples from populations whose distributions are not normal. But what happens when you have several such samples to compare? As discussed in Chapter 7 pair-wise comparisons of the various samples with each other using multiple applications of the same test is not desirable for sound statistical reasons. The **Kruskal–Wallis test** overcomes this problem by allowing us to make comparisons between any number of sample medians in a single test.

11.2 ▶ Uses

The Kruskal–Wallis test is a non-parametric alternative to one-way ANOVA. It is used to compare three or more samples, and it tests the null hypothesis that there are no significant differences between the median values of the different samples. The interpretation of the Kruskal–Wallis test is essentially similar to that of the one-way ANOVA, except that it is based on ranks rather than means. The test statistic, H, that is calculated in the Kruskal–Wallis test for k samples has a **sampling distribution** close to that of the χ^2 distribution (see Chapter 15) for k-1 degrees of freedom.

11.3 ▶ Formulae

Let there be k samples and let n_i be the number of observations in the ith sample. If $\bar{R}_i$ is the mean rank of the ith sample and $\bar{R}$ is the overall mean rank of all the observations:

11.3.1 Sum of squares of ranks (SS$_R$)

$$SS_R = \sum_{i=1}^{k} n_i \bullet \left(\bar{R}_i - \bar{R} \right)^2$$

$$H = \frac{12}{N(N+1)} SS_R$$

where $N = \sum_{i=1}^{k} n_i$

Degrees of freedom df = k-1

11.3.2 Correction for ties

$$H = \frac{SS_R \dfrac{12}{N(N+1)}}{1 - \dfrac{\sum_{j=1}^{q}[(T_j - 1)T_j(T_j + 1)]}{N(N^2 - 1)}}$$

where q is the number of pooled ranks with ties, and where T_j is the number of ties in the jth rank with ties

 Step-by-step tutorial 11.1: Kruskal–Wallis test

A biologist counted the number of beetles captured in pitfall traps in four different types of woodland with each woodland area having five traps. The results are given in the Table 11.1.

Table 11.1   Beetles captured in pit fall traps

Woodland 1	Woodland 2	Woodland 3	Woodland 4
2	8	34	13
15	12	45	5
23	18	65	32
50	60	92	19
40	22	67	25

Step 1   We first assign to each value its rank in the whole collection of data (pooled ranks). Tied ranks are assigned their average rank (Table 11.2).

Table 11.2   Beetles in traps: ranking each value

Woodland 1		Woodland 2		Woodland 3		Woodland 4	
Data	Rank	Data	Rank	Data	Rank	Data	Rank
2	1	8	3	34	13	13	5
15	6	12	4	45	15	5	2
23	10	18	7	65	18	32	12
50	16	60	17	92	20	19	8
40	14	22	9	67	19	25	11

Step 2   Calculate the mean rank $\bar{R}_i$ for each group and the overall mean rank $\bar{R}$.

$$\overline{R}_1 = (1+6+10+16+14)/5 = 9.4$$

$$\overline{R}_2 = (3+4+7+17+9)/5 = 8.0$$

$$\overline{R}_3 = (13+15+18+20+19)/5 = 17.0$$

$$\overline{R}_4 = (5+2+12+8+11)/5 = 7.6$$

$$\overline{R} = \frac{\sum i}{N} = (1+2+3+4+\ldots+17+18+19+20)/20 = 10.5$$

Step 3   Calculate the sum of squares (SS_R) of the rank deviates:

$$SS_R = \sum n_i(\overline{R}_i - \overline{R})^2 = 5\times(9.4-10.5)^2 + 5\times(8.0-10.5)^2 + 5\times(17.0-10.5)^2$$
$$+ 5\times(7.6-10.5)^2$$
$$= 5\times(1.21+6.25+42.25+8.41)$$
$$= 290.6$$

Step 4   Calculate H:

$$H = \frac{12}{N(N+1)}SS_R = \frac{12}{20\times21}290.6 = 8.303$$

Step 5   Compare H with the critical value of χ^2 for $4 - 1 = 3$ degrees of freedom (Appendix 6). $\chi^2(3; 0.05) = 7.815$. Since the calculated value of H is greater than 7.815 we reject the null hypothesis and conclude that there is a significant difference between the number of beetles trapped in the different woodlands.

    Note that this result only indicates that there are differences within the group as a whole. Caution is needed in drawing inferences about particular *pairs of samples*. However it is safe to conclude that there is, at the very least, a significant difference between the two groups with the highest and lowest mean average rank sum, in this example between Woodlands 3 and Woodlands 4.

 PractiStat tutorial 11.1: Kruskal–Wallis test

Step 1   Select the OPEN button on the toolbar or select *Open data file* from the *File* menu and navigate to the 'Samples' folder on the PractiStat CD-ROM. Open the data file 'Woodland-1'. The data from this file should now be displayed and the name 'Woodland-1' should appear on the samples list.

Step 2   Similarly open the files 'Woodland-2', 'Woodland-3' and 'Woodland-4'.

Step 3   Select *Kruskal–Wallis H-test* from the *Statistics* menu or click on the H-TEST button. Multi-select 'Woodland-1', 'Woodland-2', 'Woodland-3' and 'Woodland-4' from the samples list (Control-click on PCs or Command-click on Macs).

Step 4   The viewing area should now show the results of a Kruskal–Wallis *H*-test for these samples. The calculated value of *H* is 8.303. The program shows that for 3 degrees of freedom P < 0.05. We can hence conclude that there was a significant difference between the numbers of beetles trapped in the different woodlands.

11.4 ▶ Assumptions underlying the Kruskal–Wallis test

If there are only three samples there should be more than five observations in each sample.

12

The Friedman test

Comparing several matched samples using a non-parametric method

Chapter 12 examines how the non-parametric Friedman test can be used to compare the medians of several matched groups – when we need to determine if there is an effect of three or more treatments, for example, and so cannot use the Wilcoxon Signed-Rank test.

12.1 Description

In Chapter 7 we saw how the non-parametric Wilcoxon Signed-Rank test determined whether the median of observed differences after a given treatment deviated enough from zero to conclude that the treatment had a significant effect. When a series of three or more treatments is administered and we need to determine whether there is an effect of the treatments, the **Friedman test** may be used. The Friedman test is a non-parametric method for comparing the *medians of several matched groups*.

12.2 Uses

When one or more of the assumptions of the one-way ANOVA for matched samples (see Chapter 8) are not met the Friedman test is an appropriate alternative. Its use may be called for when the distributions of the samples are far from normal or when the data are not on equal-measurement scales – e.g. they are ranks or ratings or are on intrinsically non-linear scales. The test is similar to the Kruskal–Wallis test described in Chapter 11, except that it is used for *matched* groups.

12.3 Formulae

Let there be n subjects and k measures per subject. If $\bar{R}_i$ is the mean rank of the ith measure and $\bar{R}$ is the overall mean rank of all the observations.

12.3.1 Between-samples sum of squares

$$SS_B = n \sum_{i=1}^{k} (\bar{R}_i - \bar{R})^2$$

12.3.2 χ^2

$$\chi^2 = 12 \bullet \frac{SS_B}{k(k+1)}$$

12.3.3 Degrees of freedom

$$df = k - 1$$

Step-by-step tutorial 12.1: Friedman test

A dermatologist wishes to test the efficacy of an antileukotriene drug, montelukast, in the treatment of the skin complaint chronic idiopathic urticaria

Table 12.1 Symptomatic profile of patients

Patient #	Total symptom score		
	Before treatment	**After 15 days**	**After 30 days**
1	8	6	7
2	15	14	10
3	12	12	4
4	19	15	14
5	13	9	1
6	14	10	11
7	17	10	9
8	7	10	8
9	11	8	0
10	16	13	3

(hives). The drug was administered to 10 patients over a period of 30 days. The symptomatic profile of each patient before and during the test was evaluated by assigning a score from 0 to 3 to each of the different symptoms and obtaining a total symptom score for each patient by adding the points. The results were as in Table 12.1.

Step 1 First rank-order the total symptom scores for each patient, assigning a rank of 1 to the lowest score and a rank of 3 to the highest score. Remember to assign tied ranks their average rank. Then calculate the mean rank scores for each column as shown in Table 12.2.

Step 2 Calculate the between-samples sum of squares:

$$SS_B = n\sum_{i=1}^{k}(\bar{R}_i - \bar{R})^2 = 10\left((2.75-2)^2 + (1.95-2)^2 + (1.30-2)^2\right) = 10.55$$

Step 3 Calculate χ^2:

$$\chi^2 = 12 \bullet \frac{SS_B}{k(k+1)} = 12 \bullet \frac{10.55}{3(3+1)} = 10.55$$

Step 4 Compare the value of χ^2 with the critical value of χ^2 for $k - 1 = 2$ degrees of freedom:

$$\chi^2 = (2; 0.05) = 5.99$$

Since the calculated value is greater than the critical value we can conclude that administration of montelukast was an effective treatment for urticaria.

 PractiStat tutorial 12.1: Friedman test

Step 1 Select the OPEN button on the toolbar or select *Open data file* from the *File* menu and navigate to the 'Samples' folder on the PractiStat CD-

Table 12.2 Ranking the symptom scores

Patient #	Before treatment		After 15 days		After 30 days	
	Symptom score	Rank	Symptom score	Rank	Symptom score	Rank
1	8	3	6	1	7	2
2	15	3	14	2	10	1
3	12	2.5	12	2.5	4	1
4	19	3	15	2	14	1
5	13	3	9	2	1	1
6	14	3	10	1	11	2
7	17	3	10	2	9	1
8	7	1	10	3	8	2
9	11	3	8	2	0	1
10	16	3	13	2	3	1
Total rank	27.5		19.5		13	
Mean rank	2.75		1.95		1.30	

Overall mean rank = (1 + 2 + 3)/3 = 2

ROM. Open the data file 'Urticaria 0 days'. The data from this file should now be displayed and the name 'Urticaria 0 days' should appear on the samples list.

Step 2 Similarly open the files 'Urticaria 15 days' and 'Urticaria 30 days'.

Step 3 Select *Friedman Test* from the *Statistics* menu or click on the FRIEDM button. Multi-select 'Urticaria 0 days', 'Urticaria 15 days' and 'Urticaria 30 days' from the samples list (Control-click on PCs or Command-click on Macs).

Step 4 The viewing area should now show the results of a Friedman test. The calculated χ^2 (Chi Square) value is 10.55 and the critical value of χ^2 for $k - 1$ degrees of freedom, shown as Chi critical (2; 0.05), is 5.99. Since the calculated value of 10.55 is greater than the critical value of 5.99 we can conclude that administration of montelukast was an effective treatment for urticaria.

13

Two-way ANOVA

Comparing several groups for the effects of two independent variables

Chapter 13 introduces the concept of interaction between variables, and explains how two-way ANOVA can help us assess the effect of two independent variables on one dependent variable in our experimental design.

 Description

Two-way ANOVA allows us to assess the effect of two independent variables on one dependent variable.

 Uses

For example, we may be interested in determining whether diet (our independent variable) *and* gender (another independent variable) have an effect on weight gain in laboratory rats. We might try using the one-way ANOVA twice; first to investigate the effect of diet and then to investigate the effect of gender, but this method would fail to ascertain whether there was any **interaction** between our two independent variables. Just as important as the effect of diet and the effect of gender on weight gain is the effect of unique combinations of diet and gender on weight gain. In other words, there may be a significant effect on weight gain when *female* rats are fed only *one particular* diet and that particular diet has no effect on weight gain in male rats.

The concept of **interaction** is illustrated in Figures 13.1(a)–13.1(f).

In Figures 13.1(a), (b) and (c), there is no interaction between diet and gender. Where there are effects of diet or gender (Figures 13.1(b) and (c)), changes observed as a result of one independent variable are associated with similar changes observed as a result of the other independent variable – i.e. there is no interaction between the variables. In contrast to the first three scenarios, Figures 13.1(d), (e) and (f) show how unique combinations of diet and gender can result in an effect on weight gain. Figures 13.1(d) and (e) show how a unique combination of Diet 1 in males can result in an effect of diet when no other combination of diet and gender has a significant effect. The interactions in Figures 13.1(d) and (e) are sometimes referred to, respectively, as **positive interaction** and **negative interaction**.

In two-way ANOVA, as with the one-way ANOVA, we partition the total sum of squares into between-samples sum of squares, SS_B, and within-samples sum of squares, SS_W. However, SS_B is subdivided into three components:

1 Sum of squares for variation due to independent variable 1

2 Sum of squares for variation due to independent variable 2

3 Sum of squares for variation due to *interaction* between variables 1 and 2.

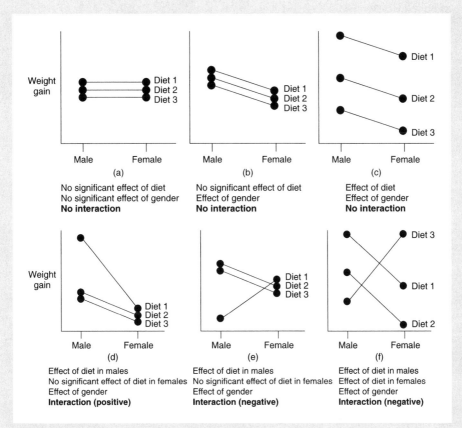

Figure 13.1 The concept of interaction

13.3 ▶ Formula

The formulae for data arranged in rows and columns with r categories of independent variable 1 and c categories of variable 2 and with n values in each group are shown in Table 13.1.

13.3.1 Overall mean

$$\overline{X} = \frac{1}{rcn} \sum_{i=1}^{r} \sum_{j=1}^{c} \sum_{k=1}^{n} x_{ijk}$$

The formulae are shown in Table 13.2.

Table 13.1 Two-way ANOVA data arrangement

		Independent variable 2			
		Category 1	...	Category c	Row means $\bar{R}$
Independent variable 1	Category 1	$x_{1,1,1}$	...	$x_{1,c,1}$	
		...	...	...	
		$x_{1,1,n}$	...	$x_{1,c,n}$	
		$\bar{x}_{1,1}$	...	$\bar{x}_{1,c}$	$\bar{R}_1$
	...	...	...	...	...
	Category r	$x_{r,1,1}$	...	$x_{r,c,1}$	
		...	...	...	
		$x_{r,1,n}$	...	$x_{r,c,n}$	
		$\bar{x}_{r,1}$	...	$\bar{x}_{r,c}$	$\bar{R}_r$
	Column means $\bar{C}$	$\bar{C}_1$	...	$\bar{C}_c$	

Table 13.2 Two-way ANOVA formulae

Source of variation	Degrees of freedom	Sum of squares	Mean square	F
Variable 1 (row variable)	$df_R = r - 1$	$\bar{R}$	$MS_R = \dfrac{SS_R}{df_R}$	$F_R = \dfrac{MS_R}{MS_W}$
Variable 2 (column variable)	$df_C = c - 1$	—	$MS_C = \dfrac{SS_C}{df_C}$	$F_C = \dfrac{MS_C}{MS_W}$
Interaction	$df_I = (r-1)(c-1)$	$SS_I = SS_B - SS_C - SS_R$	$MS_I = \dfrac{SS_I}{df_I}$	$F_I = \dfrac{MS_I}{MS_W}$
Within-samples	$df_W = rc(n-1)$	$SS_W = \sum^{rc}\sum^{n}(x - \bar{x})^2$	$MS_W = \dfrac{SS_W}{df_W}$	
Between-samples	$df_B = rc - 1$	$SS_B = n\sum^{rc}(x - \bar{X})^2$	$MS_B = \dfrac{SS_B}{df_B}$	
Total	$df_T = rcn - 1$	$SS_T = SS_W + SS_B$	$MS_T = \dfrac{SS_T}{df_T}$	

 ## Step-by-step tutorial 13.1: two-way ANOVA

We return to the data above given previously in Chapter 8 for the weight gain of male rats fed different diets. We now also carry out similar measurements on female rats. Thus both diet and gender are the two independent variables whose effect on the dependent variable, weight gain, we wish to ascertain. The data are tabulated in Table 13.3.

Let r be the number of categories of variable 1 (Gender), c be the number of categories of variable 2 (Diet) and $\bar{x}$ be the mean value for a group.

Table 13.3 Weight gain of laboratory rats

		Variable 2, diet		
		Diet 1	Diet 2	Diet 3
Variable 1, gender	Male	90	120	125
		95	125	130
		100	130	135
	Female	75	88	118
		78	95	125
		90	105	132

Step 1 First calculate the mean values as shown in Table 13.4, denoting the row means as $\bar{R}$ and the column means as $\bar{C}$.

Table 13.4 Weight gain: calculating mean values

		Variable 2, diet			
		Diet 1	Diet 2	Diet 3	Row means $\bar{R}$
Variable 1, gender	Male	90	120	125	
		95	125	130	
		100	130	135	
		Mean = 95	Mean = 125	Mean = 130	116.7
	Female	75	100	118	
		78	118	125	
		90	112	132	
		Mean = 81	Mean = 110	Mean = 125	105.3
	Column means $\bar{C}$	88	117.5	127.5	

Step 2 Calculate the overall mean $\bar{X}$:

$$\bar{X} = \frac{1}{rcn}\sum^{r}\sum^{c}\sum^{n}x$$

$$= \frac{1}{(2)(3)(3)}((90+95+100)+(120+125+130)+(125+130+135)$$

$$+ (75+78+90)+(100+118+112)+(118+125+132))$$

$$= \frac{1}{18}(1998)$$

$$= 111$$

Step 3 Calculate the between-samples sum of squares:

$$SS_B = n \sum_{}^{rc} (\bar{x} - \bar{X})^2$$
$$= 3((95 - 111)^2 + (125 - 111)^2 + (130 - 111)^2$$
$$+ (81 - 111)^2 + (110 - 111)^2 + (125 - 111)^2)$$
$$= 5730$$

Step 4 Calculate the within-samples sum of squares:

$$SS_W = \sum_{}^{rc} \sum_{}^{n} (x - \bar{X})^2$$
$$= (90 - 95)^2 + (95 - 95)^2 + (100 - 95)^2$$
$$+ (120 - 125)^2 + (125 - 125)^2 + (130 - 125)^2$$
$$+ (125 - 130)^2 + (130 - 130)^2 + (135 - 130)^2$$
$$+ (75 - 81)^2 + (78 - 81)^2 + (90 - 81)^2$$
$$+ (100 - 110)^2 + (118 - 110)^2 + (112 - 110)^2$$
$$+ (118 - 125)^2 + (125 - 125)^2 + (132 - 125)^2$$
$$= 542$$

Step 5 Calculate the total sum of squares:

$$SS_T = SS_B + SS_W$$
$$= 5730 + 542$$
$$= 6272$$

Step 6 Calculate the sum of squares due to gender:

$$SS_{Gender} = \sum_{}^{r} cn(\bar{R} - \bar{X})$$
$$= 9((116.7 - 111)^2 + (105.3 - 111)^2)$$
$$= 585$$

Step 7 Calculate the sum of squares due to diet:

$$SS_{Diet} = \sum_{}^{c} rn(\bar{C} - \bar{X})$$
$$= 6((88 - 111)^2 + (117.5 - 111)^2 + (127.5 - 111)^2)$$
$$= 5061$$

Step 8 Calculate the sum of squares due to interaction:

$$SS_{inter} = SS_B - SS_{Gender} - SS_{Diet}$$
$$= 5730 - 585 - 5061$$
$$= 84$$

Table 13.5 Weight gain: completed ANOVA table

Source of variation	Degrees of freedom	Sum of squares	Mean square	F	P
Gender (between rows)	$r - 1 = 1$	585	585	585/45.2 = 12.9	<0.01
Diet (between columns)	$c - 1 = 2$	5061	2530	2530/45.2 = 56.0	<0.01
Interaction	$(r - 1)(c - 1) = 2$	84	42	42/45.2 = 0.9	Not significant
Within-samples	$rc(n - 1) = 12$	542	45.2		
Total	$rcn - 1 = 17$	6272	369		

Step 9 Set up the completed ANOVA table (Table 13.5) and compare the calculated values of F with the critical values of F for the appropriate degrees of freedom, as given in Appendix 2.

We can conclude that both diet and gender have a significant effect on the weight gain of rats. We also can conclude that there is no significant interaction between gender and diet in weight gain.

13.4 Assumptions of two-way ANOVA

Assumptions 1–4 of the one-way ANOVA apply:

1 The scale on which the dependent variable is measured is an equal interval scale

2 The samples are drawn independently and randomly from the source population(s)

3 The source population(s) have a normal distribution

4 The samples have approximately equal variances (ratio of largest to smallest variance not more than 1.5)

plus:

5 Effects on the dependent variable due to the two independent variables should be *additive*, i.e. any effect of independent variable A should be the same for each level of independent variable B. In other words, the interaction between independent variables A and B should not be significantly different from zero. This is why the probability of interaction in the two-way ANOVA table should be significantly low before making conclusions on the effects of the independent variables.

14

Correlation and regression

Establishing relationships between two variables

Chapter 14 introduces the concept of relationship between two variables, when their values correspond to each other in some systematic way. If we wish to know the strength of any relationship we have observed, and how reliable this observation is, we need to employ a correlation coefficient and regression analysis. We first distinguish between correlation (if the variables are related and how far they move together) and regression (determining the mathematical relationship between several independent or predictor variables and a single dependent or criterion variable). We then examine Pearson's correlation coefficient, the most widely used correlation coefficient for quantitative data, and Spearman's rank correlation coefficient, a non-parametric test based on ranks which is appropriate for ordinal or ranked data. Finally we look at simple linear regression, where the data can be visualised and analysed in a scatter plot with a best straight line fitted to the data points (the regression line or least squares line), and show how to deal with any sampling error.

14.1 ▶ Description

Finding relations between variables is an important goal for the experimenter. Two variables are related if their values *correspond to each other in some systematic way*. For example, tall people are usually heavier than short people; therefore height and weight are related variables. Statistics helps us to establish the nature of the relations between variables. In particular we wish to know (a) how strong is the relationship we have observed and (b) how reliable is our observation of a relationship.

Two common measures of the relationship between variables are the **correlation coefficient** and **regression analysis**. **Correlation** is a measure of the relationship between two or more variables that help us determine whether the variables really are related and the degree to which they vary together. **Regression** is a statistical tool for determining the mathematical relationship between several independent or *predictor* variables and a single dependent or *criterion* variable, allowing us to calculate the value of one variable given known values of the other variables.

Correlation coefficients can range from −1.00 (a perfect negative correlation) to +1.00 (a perfect positive correlation). A value of 0.00 represents a complete lack of evidence of any correlation. The most widely used correlation coefficient for quantitative data is **Pearson's correlation coefficient** (also called the **product moment correlation coefficient**). **Spearman's rank correlation coefficient** is a non-parametric test based on ranks which is appropriate for ordinal or ranked data.

The simplest case of regression analysis is **linear regression** with one dependent variable and one independent variable. The data can be visualised in a **scatter-plot** and analysed by fitting the **best straight line** to the points. This line, called the **regression** line, is also called the **least squares line** because it is determined such that the sum of all the squared distances of the actual dependent values to their corresponding predicted values has the lowest possible value. The slope of the regression line is called the **regression coefficient**.

14.2 ▶ Uses

Correlation analysis allows us to determine whether two variables have a linear association and the degree to which they vary together. It is important to note that there is no distinction between independent and dependent variables and a causal relationship between the two variables is not assumed; commonly both variables are the result of a common cause. Regression analysis allows us to describe the functional relationship between two variables and predict the value of one variable in terms of the other. It is frequently assumed that there may be a cause-and-effect relationship between the two variables. However the observation of a functional relationship between two variables is not sufficient grounds for establishing that one causes the other.

For both correlation and regression the variables studied *must* be quantitative – the techniques have no application to qualitative variables.

14.3 Formulae

14.3.1 Linear regression

Regression line

$$y = a + bx$$

Regression coefficient (slope of regression line)

$$b = \frac{n\sum xy - \sum x \sum y}{n\sum x^2 - \left(\sum x\right)^2}$$

Intercept of regression line

$$a = \bar{y} - b\bar{x}$$

14.3.2 Correlation (parametric)

Pearson's correlation coefficient

$$r = \frac{n\sum xy - \sum x \sum y}{\sqrt{\left[n\sum x^2 - \left(\sum x\right)^2\right]\left[n\sum y^2 - \left(\sum y\right)^2\right]}}$$

t-test for Pearson's correlation coefficient

$$t = \frac{r}{\sqrt{\dfrac{1 - r^2}{n - 2}}}$$

Degrees of freedom for t-test

$$df = n - 2$$

14.3.3 Correlation (non-parametric)

Spearman's rank correlation coefficient

$$r_s = 1 - \frac{6\sum_{i=1}^{n} \Delta R_i^2}{n^3 - n}$$

where ΔR_i is the difference in rank between the x and y components of the ith data point

For $n < 30$ a t-test for Spearman's rank correlation coefficient is used:

$$t = \frac{r_s}{\sqrt{\dfrac{1 - r^2}{n - 2}}}$$

 Step-by-step tutorial 14.1: linear regression for estimating unknown values from a standard curve

One method for measurement of phosphate concentrations uses the dye malachite green to generate a coloured compound whose absorbance can be measured spectrophotometrically. Standard phosphate solutions are used to set up a standard curve from which unknown values can be estimated. The results obtained were as in Table 14.1 and Figure 14.1.

Step 1 The slope, b, and intercept, a, of the regression line $y = a + bx$ are calculated as in Table 14.2.

$$b = \frac{n\sum xy - \sum x \sum y}{n\sum x^2 - \left(\sum x\right)^2} = \frac{6 \times 14.39 - 26 \times 2.25}{6 \times 184 - 676} = 0.065$$

$$a = \bar{y} - b\bar{x} = 0.375 - 0.065 \times 4.33 = 0.094$$

Table 14.1 Absorbance of standard phosphate concentrations

Phosphate standard (nmol)	Absorbance
0	0.12
1	0.18
3	0.26
5	0.34
7	0.59
10	0.76

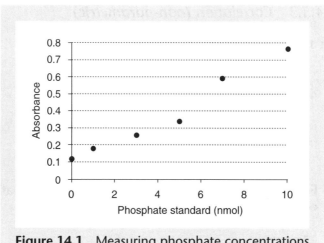

Figure 14.1 Measuring phosphate concentrations

Table 14.2 Phosphate standard curve: calculating the slope and intercept of the regression line

x	y	xy	x^2
0	0.12	0	0
1	0.18	0.18	1
3	0.26	0.78	9
5	0.34	1.70	25
7	0.59	4.13	49
10	0.76	7.6	100
$\sum x = 26$	$\sum y = 2.25$	$\sum xy = 14.39$	$\sum x^2 = 184$
$\bar{x} = 4.33$	$\bar{y} = 0.375$		
$\left(\sum x\right)^2 = 6763$			

Step 2   Hence the regression line is $y = 0.094 + 0.065x$

Step 3   Values of absorbance in test samples can now be converted to nmol phosphate by substitution in the regression line curve – e.g. an absorbance of 0.45 would correspond to

$$\frac{0.45 - 0.094}{0.065} = 5.48 \,\text{nmol phosphate.}$$

PractiStat tutorial 14.1: regression

Step 1 Select the OPEN button on the toolbar or select *Open data file* from the *File* menu and navigate to the 'Samples' folder on the PractiStat CD-ROM. Open the data file 'Phosphate absorbance'. The data from this file should now be displayed and the name 'Phosphate absorbance' should appear on the samples list.

Step 2 Similarly open the file 'Phosphate concentration'.

Step 3 Select *Linear Regression* from the *Statistics* menu or click on the REG button on the toolbar. With 'Phosphate concentration' selected in the samples list, click on 'Mark as Independent [X]'.

Step 4 Multi-select 'Phosphate absorbance' and 'Phosphate concentration' from the samples list. To select multiple samples from the list, select the first sample by clicking on it then holding down the 'CNTL' key on PCs or the Command-key on Macs while clicking on subsequent samples.

Step 5 The viewing area should now show the results of the regression analysis. The equation of the regression line is given as $y = 0.094 + 0.065x$. To change the plot parameters of the linear regression, press the 'Edit Graph ...' button, enter the new plot parameters then press the 'OK' button.

14.4 Regression and causation

If we wish to investigate the hypothesis that changes in an independent variable, X, *cause* changes in a dependent variable, Y, we may perform an experiment in which X is systematically varied and the response of Y measured. Our hypothesis would indeed be strengthened if we find that there is a statistically significant **linear relationship** of Y to X. Without straying too far into philosophical territory, however, we should note that such a finding would not *prove* our theory. Only the converse is true – i.e. a failure to find a significant regression of Y on X would, providing the experiment had been carefully controlled to exclude confounding influences, be taken to *disprove* the theory.

14.5 Curve fitting

How do we fit our bivariate data to the **best straight line** for linear regression analysis (or to a more complicated function for non-linear analysis)?

Consider the following data for two variables X and Y shown in Table 14.3 and plotted in Figure 14.2.

Figures 14.3 and 14.4 show two straight lines drawn through the data.

Table 14.3 Data for X and Y

X	Y
24	22
58	12
86	20
100	40
125	30
150	24
170	48
190	40

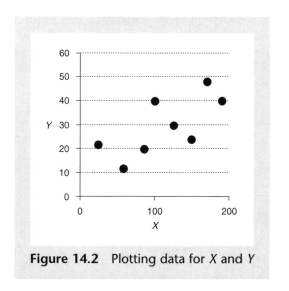

Figure 14.2 Plotting data for X and Y

How do we decide which line best fits the data? We draw vertical lines from each point to the fitted line. These vertical lines represent the deviations of the experimental values from those predicted by the fitted line (Figure 14.5).

If d_i is the deviation of the data pair $[x_i, y_i]$ from the fitted line at x_i, then the **best straight line** is that for which Σd_i^2 is a minimum. The deviations, d_i are referred to as **residuals** and Σd_i^2 is called the **residual sum of squares**. If the predicted value of y for a given value of x is written as $\hat{y}$, then $d_i = y_i - \hat{y}_i$ and the residual sum of squares is $\Sigma(y_i - \hat{y}_i)^2$.

For linear regression the equation of the best fit line can be calculated from first principles, as described below. However for non-linear regression recursive methods must be used in which initial estimates of the parameters are made and then systematically varied until Σd_i^2 shows no significant further reduction.

The equation for a straight line is $y = a + bx$, where b is the slope and a is the intercept on the y-axis. As shown in Box 14.1, the straight line for which Σd_i^2 is a minimum is given by:

$$y = \frac{\sum (x_i - \bar{x})(y_i - \bar{y})}{\sum (x_i - \bar{x})^2} x + \bar{y} - \frac{\sum (x_i - \bar{x})(y_i - \bar{y})}{\sum (x_i - \bar{x})^2} \bar{x}$$

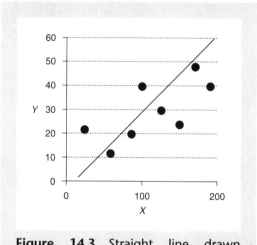

Figure 14.3 Straight line drawn through data for X and Y:1

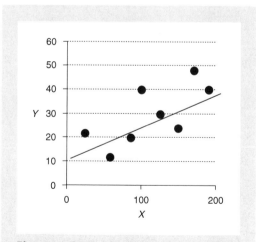

Figure 14.4 Straight line drawn through data for X and Y:2

Notice that when $x = \bar{x}$, $y = \bar{y}$ – i.e. the regression line passes through the point $(\bar{x}, \bar{y})$.

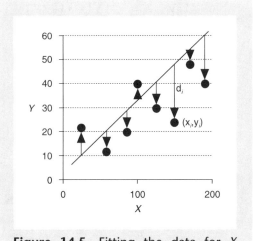

Figure 14.5 Fitting the data for X and Y

14.6 How good a fit?

Since the regression line is derived from sample data it is subject to **sampling error**. A measure of how closely the data points cluster round the fitted line is given by the residual standard deviation, denoted $s_{y\backslash x}$:

$$s_{y\backslash x} = \sqrt{\frac{\sum d_i^2}{n-2}}$$

If the residuals are normally distributed then we expect 68% of the data points to be within a vertical distance of ±1 $s_{y\backslash x}$ from the regression line and 95% to be within ±2$s_{y\backslash x}$.

The standard error of the regression coefficient, SE_{rc}, is given by the equation:

$$SE_{rc} = \frac{s_{y\backslash x}}{\sqrt{\sum (x_i - \bar{x})^2}}$$

14.7 What about curved relationships?

Simple linear regression depends on there being a *linear relation* between the variables. Often, however, the raw data exhibit a non-linear relation. In such cases transformation of the x- and/or y-data may result in a linear relationship. Table 14.4 and Figure 14.6 illustrate this procedure.

A plot of x vs y is clearly curved and linear regression cannot be applied. However if the y-data are transformed by taking logarithms, the plot of

Table 14.4 Non-linear data

X	Y
2	23
4	45
5	67
8	80
8.7	150
10	278
12	500

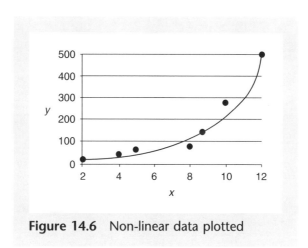

Figure 14.6 Non-linear data plotted

$\log(y)$ vs x is now linear and simple regression is possible (Table 14.5 and Figure 14.7).

For some data sets logarithmic transformation of both axes may be required to achieve a linear relationship.

Step-by-step tutorial 14.2: Pearson's correlation

A dermatologist was studying the abundance in breast tissue of a type of cell known as a Langerhans cell. In breast biopsies from 12 healthy women he recorded the number of Langerhans cells and the number of other epithelial cells per mm^2. The results are given in Table 14.6.

The correlation coefficient is calculated as follows.

Step 1 Calculate Σx, Σy, Σxy, Σx^2, Σy^2. These values are tabulated in Table 14.7.

Step 2 Calculate Pearsons's correlation coefficient, r, as:

Table 14.5 Logarithmic transformation of y-data

x	y	log(y)
2	23	1.362
4	45	1.653
5	67	1.826
8	80	1.903
8.7	150	2.176
10	278	2.444
12	500	2.699

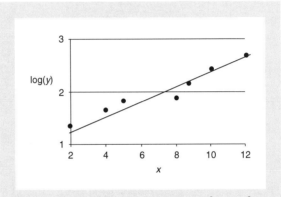

Figure 14.7 Linear regression of transformation med-data

Table 14.6 Epithelial cell types

Subject #	No. of cells per mm²	
	Langerhans cells	Other cells
1	912	47315
2	1023	50234
3	1345	58902
4	1002	59078
5	1458	68078
6	1560	76089
7	1578	85076
8	1678	78034
9	1409	93456
10	1378	90236
11	1709	99789
12	1806	104580

Table 14.7 Langerhans and other cells: calculating the correlation coefficient

Subject	No. of cells per mm²		xy	x^2	y^2
	Langerhans cells (x)	Other cells (y)			
1	912	47315	4.315×10^7	8.317×10^5	2.238×10^9
2	1023	50234	5.139×10^7	1.047×10^6	2.523×10^9
3	1345	58902	7.922×10^7	1.809×10^6	3.469×10^9
4	1002	59078	5.920×10^7	1.004×10^6	3.490×10^9
5	1458	68078	9.926×10^7	2.126×10^6	4.635×10^9
6	1560	76089	1.187×10^7	2.434×10^6	5.790×10^9
7	1578	85076	1.342×10^8	2.490×10^6	7.238×10^9
8	1678	78034	1.309×10^8	2.816×10^6	6.089×10^9
9	1409	93456	1.317×10^8	1.985×10^6	8.734×10^9
10	1378	90236	1.243×10^8	1.899×10^6	8.143×10^9
11	1709	99789	1.705×10^8	2.921×10^6	9.958×10^9
12	1806	104580	1.889×10^8	3.262×10^6	1.094×10^{10}
	$\sum x =$ 16858	$\sum y =$ 910867	$\sum xy =$ 1.331×10^9	$\sum x^2 =$ 2.462×10^7	$\sum y^2 =$ 7.324×10^{10}
	$\left(\sum x\right)^2 =$ 2.842×10^8	$\left(\sum y\right)^2 =$ 8.297×10^{11}			

$$r = \frac{n\sum xy - \sum x \sum y}{\sqrt{\left[n\sum x^2 - \left(\sum x\right)^2\right]\left[n\sum y^2 - \left(\sum y\right)^2\right]}}$$

$$= \frac{12 \times 1.331 \times 10^9 - 16858 \times 910867}{\sqrt{\left[12 \times 2.462 \times 10^7 - 2.842 \times 10^8\right]\left[12 \times 7.324 \times 10^{10} - 8.297 \times 10^{11}\right]}}$$

$$= 0.829$$

The dermatologist concluded that the number of Langerhans cells is not random but is related in a linear manner to the concentration of non-Langerhans epithelial cells.

Step 3 To assess the significance of the correlation, calculate t from:

$$t = \frac{r}{\sqrt{\dfrac{1-r^2}{n-2}}}$$

$$= \frac{0.836}{\sqrt{\dfrac{1-0.836^2}{12-2}}}$$

$$= 4.815$$

From Appendix 1, the critical value of *t* at P < 0.01 for 12 − 2 = 10 degrees of freedom is 3.169. Therefore the correlation is significant at P < 0.01.

 ## PractiStat tutorial 14.2: Pearson's correlation

Step 1 Select the OPEN button on the toolbar or select *Open data file* from the *File* menu and navigate to the 'Samples' folder on the PractiStat CD-ROM. Open the file titled 'Cells Langerhans'. The data from this file should now be displayed and the name 'Cells Langerhans' should appear on the samples list.

Step 2 Similarly open the file 'Cells non-Langerhans'.

Step 3 Select *Pearson's Correlation* from the *Statistics* menu or click on the PEARSON button on the toolbar. Multi-select 'Cells Langerhans' and 'Cells non-Langerhans' from the samples list (Control-click on PCs or Command-click on Macs).

Step 4 The viewing area should now show the results of a Pearson product moment correlation analysis. The calculated value of *r* is 0.836 and of *t* is 4.815. The *t*-critical at 10 degrees of freedom at the 0.05 significance level is 2.228. The P-value is listed as '<0.05'. Therefore we can conclude that the number of Langerhans cells is related linearly to the number of non-Langerhans cells.

Step 5 To determine the ratio of non-Langerhans to Langerhans cells select the REG button on the toolbar. With 'Cells non-Langerhans' selected click on 'Select independent variable [x]'. Multi-select 'Cells Langerhans' and 'Cells non-Langerhans' from the samples list.

Step 6 The viewing area should now show the regression of number of non-Langerhans cells, *y*, on number of Langerhans cells, *x*. The equation of the regression line is given as:

$$y = 1679.784 + 55.227 \times x$$

The gradient of the line is the ratio of non-Langerhans cells to Langerhans cells – i.e. 55.227.

 ## Step-by-step tutorial 14.3: regression analysis

The data in Step-by-step tutorial 14.2 suggest that the number of Langerhans cells in breast tissue is related linearly to the number of other epithelial cell types. To calculate the ratio of Langerhans cells to other cells we can carry out linear regression.

For the data of Step-by-step tutorial 14.2 we calculate the regression coefficient, *b*, as:

$$b = \frac{n\sum xy - \sum x \sum y}{n\sum x^2 - \left(\sum x\right)^2}$$

$$= \frac{12 \times 1.331 \times 10^9 - 16858 \times 910867}{12 \times 2.462 \times 10^7 - 2.842 \times 10^8}$$

$$= 55.23$$

The intercept, a, of the regression line is given by $a = \bar{y} - b\bar{x}$

i.e. $a = \dfrac{910867}{12} - 55.23 \times \dfrac{16858}{12} = -1683.36$

The dermatologist concluded that, on average, 1 Langerhans cell is present for every 55 non-Langerhans cell.

To determine the reliability of this estimate we calculate the standard error of the regression coefficient as follows.

Step 1 First calculate the residual standard deviation, $s_{y\backslash x}$, from:

$$s_{y\backslash x} = \sqrt{\frac{\sum d_i^2}{n-2}}$$

For the epithelial cell data the values of $d_i = y_i - \hat{y}$, $d_i^2 = (y_i - \hat{y})^2$, $(x_i - \bar{x})$ and $(x_i - \bar{x})^2$ are given in Table 14.8.

Hence:

$$s_{y\backslash x} = \sqrt{\frac{\sum d_i^2}{n-2}} = \sqrt{\frac{1.237 \times 10^9}{10}} = 11121.61$$

The standard error of the regression coefficient, SE_{rc}, is given by the equation:

Table 14.8 Variable values for epithelial cell data

#	x_i	$x_i - \bar{x}$	$(x_i - \bar{x})^2$	y_i	$\hat{y}_i = 55.23x_i - 1680$	$y_i - \hat{y}_i$	$(y_i - \hat{y}_i)^2$
1	912	−492.83	2.429×10^5	47315	48687	−1372	1.883×10^6
2	1023	−381.83	1.458×10^5	50234	54817	−4583	2.101×10^7
3	1345	−59.83	3.580×10^3	58902	72600	−13698	1.876×10^8
4	1002	−402.83	1.623×10^5	59078	53657	5420	2.938×10^7
5	1458	53.17	2.827×10^3	68078	78841	−10763	1.158×10^8
6	1560	155.17	2.408×10^4	76089	84474	−8385	7.031×10^7
7	1578	173.17	2.999×10^4	85076	85468	−392	1.540×10^5
8	1678	273.17	7.462×10^4	78034	90991	−12957	1.679×10^8
9	1409	4.17	17.36	93456	76135	17320	3.000×10^8
10	1378	−26.83	720.0	90236	74423	15812	2.501×10^8
11	1709	304.17	9.252×10^4	99789	92703	7085	5.021×10^7
12	1806	401.17	1.609×10^4	104580	98060	6519	4.251×10^7
	$\bar{x} = 1404.83$		$\sum(x_i - \bar{X})^2 = 9.402 \times 10^5$			$\sum(y_i - \hat{y}_i)^2 = 1.237 \times 10^9$	

$$SE_{rc} = \frac{S_{y\setminus x}}{\sqrt{\sum (x_i - \bar{x})^2}}$$

Hence:

$$SE_{rc} = \frac{1121.62}{\sqrt{9.402 \times 10^5}} = 11.47$$

Step-by-step tutorial 14.4: Spearman's rank correlation

An ecologist is studying the effect of pesticide application on wild flower diversity. A number of fields were chosen and, where known, the number of years that had elapsed since application of pesticide was noted. The biodiversity of wild flowers was measured for each field. The scale used for biodiversity was non-linear and the records for pesticide application were incomplete, as indicated in Table 14.9. Both variables can be ranked and therefore Spearman's rank correlation coefficient analysis is appropriate. The original data and the ranks of each variable are given in Table 14.9 (the fields have been ranked in order of increasing time since application of pesticide).

The Spearman's rank correlation coefficient

$$r_s = 1 - \frac{6 \sum_{i=1}^{n} \Delta R_i^2}{n^3 - n}$$

$$r_s = 1 - \frac{6 \times 8.5}{8^3 - 8} = 0.899$$

It can be concluded that wild flower biodiversity is highly correlated with the time that has elapsed since application of pesticide. The P-value is found from the table in Appendix 7. The critical value of r_s at $P < 0.05$ for 8 obser-

Table 14.9 Pesticides and wildlife diversity: original data and variable rankings

Field #	Years without pesticides (x)	Wild flower diversity (y)	x-rank	y-rank	ΔR	ΔR^2
1	0	0.2	1	1	0	0
2	1	0.4	2	2	0	0
3	2	0.8	3	4	−1	1
4	4	0.6	4	3	1	1
5	6–8	1.4	5	7	−2	4
6	12	0.9	6.5	5	1.5	2.25
7	12	1.1	6.5	6	0.5	0.25
8	>20	1.7	8	8	0	0
						$\sum \Delta R^2 = 8.5$

vations is 0.738. Since the calculated value of r_s is greater than the critical value we conclude that the correlation is significant at $P < 0.05$.

 ### PractiStat tutorial 14.3: Spearman's rank correlation

Step 1 Select the OPEN button on the toolbar or select *Open data file* from the *File* menu and navigate to the 'Samples' folder on the PractiStat CD-ROM. Open the file titled 'Pesticide absence'. The data from this file should now be displayed and the name 'Pesticide absence' should appear on the samples list.

Step 2 Similarly open the file 'Pesticide biodiversity'.

Step 3 Select *Spearman's Correlation* from the *Statistics* menu or click on the SPEARMAN button on the toolbar. Multi-select 'Pesticide absence' and 'Pesticide biodiversity' from the samples list (Control-click on PCs or Command-click on Macs).

Step 4 The viewing area should now show the results of Spearman's rank correlation analysis. The calculated value of r_s is 0.899. The P-value is listed as '<0.05'. Therefore we can conclude that the biodiversity of wild flowers is correlated with the time elapsing since application of pesticide.

Note that PractiStat uses the *t*-test approximation for $n > 30$.

 ## 14.8 Assumptions underlying a Spearman's rank correlation analysis

For a 2-tailed test there must be at least 6 data pairs.

Box 14.1 Derivation of the regression equation for the best fit to a straight line

For a quadratic equation $y = a + bx + cx^2$ the minimum value of y is given when $\frac{\partial y}{\partial x} = 0$ i.e. when $b + 2cx = 0$

i.e. when $x = -\frac{b}{2c}$. .... *Result 1*

If the line of best fit to a set of n data pairs (x_i, y_i) is $y = a + bx$ then the deviation, d_i, of the ith observation will be given by $d_i = y_i - (bx_i + a)$

Hence:

$$\sum d_i^2 = \sum (y_i - (bx_i + a))^2 = \sum ((y_i - bx_i) - a)^2$$
$$= \sum (y_i - bx_i)^2 - 2a\sum (y_i - bx_i) + na^2$$

Continued

Box 14.1 *Continued*

From *Result 1* the minimum value of this function in *a* will occur when:

$$a = -\frac{-2\sum(y_i - bx_i)}{2n} = \frac{\sum y_i}{n} - b\frac{\sum x_i}{n}$$

i.e. when $a = \bar{y} - b\bar{x}$

Substituting this value into the regression equation gives:

$$\sum d_i^2 = \sum(y_i - (\bar{y} - b\bar{x}) - bx_i)^2$$
$$= \sum(y_i - \bar{y} + b\bar{x} - bx_i)^2$$
$$= \sum((y_i - y) - b(x_i - \bar{x}))^2$$
$$= \sum\left((y_i - \bar{y})^2 - 2b(x_i - \bar{x})(y_i - \bar{y}) + b^2(x_i - \bar{x})^2\right)$$
$$= \sum(y_i - \bar{y})^2 - 2b\sum(x_i - \bar{x})(y_i - \bar{y}) + b^2\sum(x_i - \bar{x})^2$$

This function in *b* will have a minimum when:

$$b = -\frac{-2\sum(x_i - \bar{x})(y_i - \bar{y})}{2\sum(x_i - \bar{x})}$$

i.e. when

$$b = \frac{\sum(x_i - \bar{x})(y_i - \bar{y})}{\sum(x_i - \bar{x})^2}$$

Hence the line of best fit is:

$$y = \frac{\sum(x_i - \bar{x})(y_i - \bar{y})}{\sum(x_i - \bar{x})^2}x + \bar{y} - \frac{\sum(x_i - \bar{x})(y_i - \bar{y})}{\sum(x_i - \bar{x})^2}\bar{x}$$

An alternative formulation for m can be derived as follows:

$$b = \frac{\sum(x_i - \bar{x})(y_i - \bar{y})}{\sum(x_i - \bar{x})^2} = \frac{\sum(x_i y_i - \bar{x}y_i - \bar{y}x_i + \bar{x}\bar{y})}{\sum(x_i^2 - 2\bar{x}x_i + \bar{x}^2)}$$

$$= \frac{\sum x_i y_i - \bar{x}\sum y_i - \bar{y}\sum x_i + n\bar{x}\bar{y}}{\sum x_i^2 - 2\bar{x}\sum x_i + n\bar{x}^2}$$

$$= \frac{\sum x_i y_i - \dfrac{\sum x_i}{n}\sum y_i - \dfrac{\sum y_i}{n}\sum x_i + n\dfrac{\sum x_i}{n}\dfrac{\sum y_i}{n}}{\sum x_i^2 - \dfrac{2(\sum x_i^2)}{n} + n\dfrac{(\sum x_i^2)}{n}}$$

$$\therefore b = \frac{n\sum x_i y_i - \sum x_i \sum y_i}{n\sum x_i^2 - (\sum x_i)^2}$$

15

The Chi-square tests
Hypothesis testing for categorical variables

Chapter 15 examines the use of Chi-square (χ) tests for hypothesis testing of categorical data which record the frequency of occurrence of particular events in one or more samples. We first look at the goodness of fit procedure to test whether the frequency distribution of a categorical variable is consistent with a particular theoretical distribution. We then show how Chi-square tests can be used with 2×2 or higher-order contingency tables to test for association between the variables. Finally we show how to use Yates' correction to avoid over-large Type I errors.

15.1 Description

Chi-square (χ^2) tests are used for hypothesis testing of categorical data which record the *frequency of occurrence* of particular events in one or more samples. Examples of such categorical variables include blood group, colour and gender.

15.2 Uses

χ^2 tests can be used in two main ways. First in a procedure also known as a **goodness of fit test** we can use a χ^2 test to examine whether the frequency distribution of a categorical variable is consistent with a particular theoretical distribution. For example, a biologist may use a χ^2 test to assess whether the **observed frequency** distribution of phenotypes in the progeny of a cross is consistent the **expected frequency** based on a particular model of inheritance. In this case the categorical data represent a single sample. When the data contain two or more samples – i.e. we have recorded the frequencies of categories for more than one variable – the data are arranged in **contingency tables** and a χ^2 test can be used to test for association between the variables. The simplest form of contingency table is the 2×2 table where two variables are 'crossed,' and each variable has only 2 distinct values. For example a geneticist may tabulate the frequency of 2 alleles of a gene in control subjects and subjects with a certain disease to investigate the association of the disease with a particular allele. Higher-order contingency tables can also be analysed by a χ^2 test.

15.3 Formulae

15.3.1 Goodness of fit test with m categories

$$\chi^2 = \sum_{i=1}^{m} \frac{(O_i - E_i)^2}{E_i}$$

where O_i = the observed and E_i = the expected number of occurrences for the ith category:

Degrees of freedom df = m − 1

15.3.2 2 × 2 contingency table

		Variable 1	
		Category 1	**Category 2**
Variable 2	Category 1	a	b
	Category 2	c	d

$$\chi^2 = \sum \frac{(O-E)^2}{E}$$

Degrees of freedom = 1

The following more easily calculated formula (see Box 15.1) can be used:

$$\chi^2 = \frac{(a+b+c+d)(ad-bc)^2}{(a+b)(c+d)(a+c)(b+d)}$$

15.3.3 r × c contingency table (r rows and c columns)

$$\text{Expected value } E = \frac{\text{(Row total)(Column total)}}{\text{(Grand total)}}$$

Degrees of freedom = $(r-1)\,(c-1)$

Step-by-step tutorial 15.1: goodness of fit test

A biologist counted the number of red-, white- and pink-flowered plants resulting after cross-pollination of white and red sweet peas. The results were as follows.

- **Red**: 72 plants
- **White**: 63 plants
- **Pink**: 125 plants

Mendelian inheritance of this trait predicts that the ratio of red to white to pink should be 1:1:2. Do the experimental results support this mode of inheritance?

Step 1 Calculate the values expected for the 3 colours if the inheritance does follow Mendelian genetics. The total number of plants is 72 + 63 + 125 = 260. We would therefore predict that 1/4 are red, 1/4 white and 1/2 pink – i.e. expected values are:

Red: 260/4 = 65   **White**: 260/4 = 65   **Pink**: 260/2 = 130

Table 15.1 Flower colour: observed and expected values

Colour	Observed (O)	Expected (E)	(O – E)	(O – E)²
Red	72	65	7	49
White	63	65	–2	4
Pink	125	130	–5	25

Table 15.1 tabulates the observed (O) and expected (E) values together with the calculated values of $(O - E)$ and $(O - E)^2$.

Step 2   Calculate $\chi^2 = \dfrac{49}{65} + \dfrac{4}{65} + \dfrac{25}{130} = 1.008$

Step 3   Calculate the degrees of freedom: $d.f. = m - 1 = 3 - 1 = 2$

Step 4   Compare the calculated value of χ^2 with the critical value of χ^2 with 2 degrees of freedom at $\alpha = 0.05$. χ^2 (0.05, 2) = 5.99 (Appendix 6). Since our value of χ^2 is lower than the critical value we conclude that there is no reason to reject the null hypothesis – i.e. we conclude that the observed values do not differ significantly from those predicted by the Mendelian model.

 ## PractiStat tutorial 15.1: goodness of fit test

Step 1   Select the OPEN button on the toolbar or select *Open data file* from the *File* menu and navigate to the 'Samples' folder on the BioStat CD-ROM. Open the data file 'Mendelian Observed'. The data from this file should now be displayed and the name 'Mendelian observed' should appear on the samples list.

Step 2   Similarly open the file 'Mendelian Expected'. Note the data entry format for this test (Table 15.2).

Table 15.2 Data entry format: goodness of fit test

Mendelian observed	Mendelian expected
72	65
63	65
125	130

Step 3   Select *Chi-square (goodness of fit)* from the *Statistics* menu or click on the CHI button.

Step 4   Select 'Mendelian observed' in the samples list and click on the 'Mark as observed' button.

Step 5 Multi-select 'Mendelian observed' and 'Mendelian expected' from the samples list (Control-click on PCs or Command-click on Macs).

Step 6 The viewing area should now show the results of a χ^2 goodness of fit test for these data. For the value of χ^2 calculated, 1.008, and with 2 degrees of freedom the program shows that P is not significant.

 ## Step-by-step tutorial 15.2: 2 × 2 contingency table

A neurologist measured the frequency of occurrence of antibodies to the enzyme glutamic acid decarboxylase (GAD) in the plasma of normal control subjects and of subjects with the autoimmune disease stiff-man syndrome to see whether the occurrence of GAD antibodies was associated with the disorder. A total of 550 subjects (370 subjects with stiff-man syndrome and 180 control subjects) were tested and the data are summarised in Table 15.3.

Table 15.3 Occurrence of antibodies

Subject	GAD antibody positive	GAD antibody negative
Normal	55	125
Stiff-man syndrome	220	150

Step 1 Calculate:

$$\chi^2 = \sum \frac{(O - E)^2}{E} = \frac{(a + b + c + d)(ad - bc)^2}{(a + b)(c + d)(a + c)(b + d)}$$

$$= \frac{(55 + 125 + 220 + 150)(55 \times 150 - 125 \times 220)^2}{(55 + 125)(220 + 150)(55 + 220)(125 + 150)}$$

$$= \frac{550 \times 3.706 \times 10^8}{5.036 \times 10^9}$$

$$= 40.5$$

Step 2 Degrees of freedom (df) for a 2 × 2 contingency table: $df = 1$.

Step 3 Compare the calculated value of χ^2 with the critical value of χ^2 with 1 degrees of freedom at $\alpha = 0.05$; χ^2 (0.05, 1) = 3.841 (Appendix 6). Since our value of χ^2 is greater than the critical value we reject the null hypothesis that the variables are independent and conclude that there is a significant association between the presence of GAD antibodies and the occurrence of stiff-man syndrome.

 ## PractiStat tutorial 15.2: 2 × 2 contingency table

Step 1 Select the OPEN button on the toolbar or select *Open data file* from the *File* menu and navigate to the 'Samples' folder on the PractiStat CD-ROM. Open the data file 'GAD positive'. The data from this file should

Table 15.4 Data entry format: 2 × 2 contingency table

Subject	GAD positive	GAD negative
Normal	55	125
Stiff-man syndrome	220	150

now be displayed and the name 'GAD positive' should appear on the samples list.

Step 2 Similarly open the file 'GAD negative'. Note the data entry format for this test (Table 15.4).

Step 3 Select *Chi-square Contingency Table* from the *Statistics* menu or click on the 'r × c' button.

Step 4 Multi-select 'GAD positive' and 'GAD negative' from the samples list (Control-click on PCs or Command-click on Macs).

Step 5 The viewing area should now show the results of a χ^2 2 × 2 contingency test for these data. For the value of χ^2 calculated, 40.465, the program shows that P < 0.05. We conclude that there is a significant association between the 2 variables.

15.4 Yates' correction

There is only 1 degree of freedom in a χ^2 test under 2 conditions; (i) when there are only 2 categories (e.g. male/female) in a goodness of fit test; (ii) in a 2 × 2 contingency test. Under either of these conditions the calculated value of the test statistic is too high, particularly for small samples, and results in Type I errors at a level greater than the intended one. To avoid this, **Yates' correction** is often applied to the data. This involves subtracting 0.5 from the absolute value of each 'observed–expected' term before squaring it.

Thus for a 2 × 2 contingency table the value of χ^2 when Yates' correction is applied becomes:

$$\chi^2 = \sum \frac{(|O - E| - 0.5)^2}{E} = \frac{(a+b+c+d)(|ad-bc| - 0.5(a+b+c+d))^2}{(a+b)(c+d)(a+c)(b+d)}$$

For the data of the example above on GAD antibodies, applying Yates' correction results in a value of $\chi^2 = 41.63$.

The need for Yates' correction arises from the fact that when we map discrete data (integers) to a continuous distribution (χ^2) we introduce an error owing to the 'missing' fractional part in the data. To compensate for this, we subtract 0.5. However it should be noted that some statisticians believe that use of Yates' correction makes the analysis too conservative.

 Step-by-step tutorial 15.3: higher-order contingency table

A geneticist is studying candidate genes for type 2 diabetes. A point mutation (G to A) in a potassium-channel gene has been identified which results in substitution of a lysine (K) for a glutamic acid (E) in the protein. Genotyping was carried out for this mutation in normal control subjects and in a group of randomly selected diabetic patients. The frequencies of the 3 possible genotypes (E/E, E/K and K/K) are given in Table 15.5.

Table 15.5 Potassium channel gene: frequencies of possible genotypes

Genotype	Diabetics	Controls
G/G (E/E)	133	215
G/A (E/K)	161	152
A/A (K/K)	66	30

Step 1 Calculate the row and column totals (Table 15.6).

Table 15.6 Potassium channel gene analysis: calculating row and column totals

Genotype	Diabetics	Controls	Totals
G/G (E/E)	133	215	348
G/A (E/K)	161	152	313
A/A (K/K)	66	30	96
Totals	360	397	757

Step 2 Each expected value is calculated from the formula.

$$Expected\ value\ E = \frac{(\text{Row total})(\text{Column total})}{(\text{Grand total})}$$

For example, $E\,(\text{Diabetics} - G/G) = \dfrac{(348)(360)}{(757)} = 165.5$

Step 3 The calculated values of E, $(E - O)$, $(E - O)^2$ and $\dfrac{(E - O)^2}{E}$ are shown in Table 15.7.

Step 4 We can now calculate:

Table 15.7 Potassium channel genotype frequencies: observed and expected values

Genotype	Diabetics				Controls			
	O	E	$O - E$	$\dfrac{(O - E)^2}{E}$	O	E	$O - E$	$\dfrac{(O - E)^2}{E}$
G/G (E/E)	133	165.50	−32.50	6.38	215	182.50	32.50	5.79
G/A (E/K)	161	148.85	12.15	0.99	152	164.15	−12.15	0.90
A/A (K/K)	66	45.66	20.34	9.06	30	50.35	−20.35	8.23

$$\chi^2 = \sum \frac{(O - E)^2}{E}$$
$$= 6.38 + 0.99 + 9.06 + 5.79 + 0.90 + 8.23$$
$$= 31.35$$

Step 5 Degrees of freedom $df = (r - 1)(c - 1) = (3 - 1)(2 - 1) = 2$

Step 6 The critical value of χ^2 at $\alpha = 0.05$ for 2 degrees of freedom is 5.999. Since the calculated value of χ^2 is greater than the critical value we can conclude that there is a significant association of the genotype at this locus with diabetes.

 PractiStat tutorial 15.3: higher-order contingency table

Step 1 Select the OPEN button on the toolbar or select *Open data file* from the *File* menu and navigate to the 'Samples' folder on the BioStat CD-ROM. Open the data file 'K-channel control'. The data from this file should now be displayed and the name 'K-channel control' should appear on the samples list.

Step 2 Similarly open the file 'K-channel diabetic'. Note the data entry format for this test (Table 15.8).

Table 15.8 Data entry format: higher-order contingency table

K-channel control	K-channel diabetic
215	133
152	161
30	66

Step 3 Select *Chi-square Contingency Table* from the *Statistics* menu or click on the 'r × c' button.

Step 4 Multi-select 'K-channel control' and 'K-channel diabetic' from the samples list (Control-click on PCs or Command-click on Macs).

Step 5 The viewing area should now show the results of a χ^2 contingency table for these data. For the value of χ^2 calculated, 31.347, the program shows that P < 0.05. We conclude that there is a significant association of the genotype at this locus with diabetes.

15.5 Assumptions underlying the χ^2 Chi-square tests

1 Data must be in raw counts, not proportions (fractions) or percentages.

2 An occurrence counted in one sample cannot be counted in any other sample.

3 The expected values should be relatively large. No *expected* value should be less than 1 (it does not matter what the *observed* values are) AND not more than one-fifth of the expected values should be less than 5.

Box 15.1 Derivation of the more easily calculated form of the equation for a 2 × 2 contingency table

On the null hypothesis that there is no association between the variables the expected values for each category can be derived by simple proportion as indicated in the following table:

	Observed (O)		Total	Expected (E)	
	a	b	$a + b$	$\dfrac{(a+b)(a+c)}{+b+c+d}$	$\dfrac{(a+b)(b+d)}{+b+c+d}$
	c	d	$c + d$	$\dfrac{(c+d)(a+c)}{+b+c+d}$	$\dfrac{(c+d)(b+d)}{+b+c+d}$
Total	$a + c$	$b + d$	$a + b + c + d$		

The values of $\dfrac{(O-E)^2}{E}$ are calculated as follows:

Continued

Box 15.1 Continued

Observed − Expected

$$a - \frac{(a+b)(a+c)}{a+b+c+d} = \frac{ad - bc}{a+b+c+d} \qquad\qquad b - \frac{(a+b)(b+d)}{a+b+c+d} = \frac{bc - ad}{a+b+c+d}$$

$$c - \frac{(c+d)(a+c)}{a+b+c+d} = \frac{bc - ad}{a+b+c+d} \qquad\qquad d - \frac{(c+d)(b+d)}{a+b+c+d} = \frac{ad - bc}{a+b+c+d}$$

$$\frac{(O - E)^2}{E}$$

$$\left(\frac{ad - bc}{a+b+c+d}\right)^2 \bullet \frac{1}{(a+b)(a+c)} \qquad\qquad \left(\frac{ad - bc}{a+b+c+d}\right)^2 \bullet \frac{1}{(a+b)(b+d)}$$

$$\left(\frac{ad - bc}{a+b+c+d}\right)^2 \bullet \frac{1}{(c+d)(b+d)}$$

Hence:

$$\chi^2 = \frac{(ad - bc)^2}{(a+b+c+d)} \bullet \left(\frac{1}{(a+b)(a+c)} + \frac{1}{(a+b)(b+d)} + \frac{1}{(c+d)(a+c)} + \frac{1}{(c+d)(b+d)}\right)$$

$$= \frac{(ad - bc)^2}{(a+b+c+d)} \bullet \frac{(a+b+c+d)^2}{(a+b)(a+c)(b+d)(c+d)}$$

$$= \frac{(a+b+c+d)(ad - bc)^2}{(a+b)(a+c)(b+d)(c+d)}$$

16

Useful statistics

Combining sems and calculating standard errors of differences, sums, ratios and products

Chapter 16 examines some other useful statistics for experimental designs. We first show how mean ± sem values for a particular variable can be combined to obtain an overall mean ± sem. We then show how the standard error of the difference or sum from the individual values of the sem and n for p and q can be calculated. We finally show how the standard errors of the ratio or product can be calculated from the values of the mean, sem and n for p and q.

16.1 Combining means and standard errors

16.1.1 Uses

When an experiment is repeated several times a mean ± sem value for a particular variable will be obtained for each experiment. It is useful to be able to combine these values to obtain an overall mean ± sem. This is performed as follows.

16.1.2 Formulae

For a set of k values of $x_i \pm sem_i$ $(n = n_i)$:

$$\text{Combined mean, } cmean = \frac{\displaystyle\sum_{i=1}^{k} n_i \bar{x}_i}{\displaystyle\sum_{i=1}^{k} n_i}$$

$$\text{Combined sem, } csem = \sqrt{\frac{\displaystyle\sum_{i=1}^{k}\left(n_i\left((n_i-1)sem_i^2 + \bar{x}_i^2\right)\right) - \dfrac{\left(\displaystyle\sum n_i x_i\right)^2}{\displaystyle\sum_{i=1}^{k} n_i}}{\displaystyle\sum_{i=1}^{k} n_i\left(\displaystyle\sum_{i=1}^{k} n_i - 1\right)}}$$

$$\text{Combined number of observations} = \sum_{i=1}^{k} n_i$$

 Step-by-step tutorial 16.1: combining means and sems

A biochemist carries out two separate but similar experiments measuring the rate of glucose production by liver cells. The results are given in Table 16.1.

Table 16.1 Rates of glucose production by liver cells			
Expt #	n	Glucose production ($\mu mol \cdot g^{-1} \cdot h^{-1}$)	
		Mean	sem
1	5	4.177	0.198
2	6	5.023	0.257

Step 1 The combined mean ± sem for 5 + 6 = 11 observations is calculated as follows:

$$\text{Combined mean} = \frac{5 \times 4.177 + 6 \times 5.023}{5+6}$$
$$= 4.638 \,\mu\text{mol.g}^{-1}.\text{h}^{-1}$$

Combined sem =

$$\sqrt{\frac{5 \times (0.197^2 \times 4 + 4.177^2) + 6 \times (0.257^2 \times 5 + 5.023^2) - \dfrac{(5 \times 4.177 + 6 \times 5.023)^2}{(5+6)}}{11 \times 10}}$$

$$= 0.207 \,\mu\text{mol.g}^{-1}.\text{h}^{-1}$$

Step 2 Hence the two experiments give a combined result of 4.638 ± 0.207 μmol.g^{-1}.h^{-1} (n = 11).

PractiStat tutorial 16.1: combining means and sems

Step 1 Select the OPEN button on the toolbar or select *Open data file* from the *File* menu and navigate to the 'Samples' folder on the PractiStat CD-ROM. Open the file titled 'Liver-1'. The data from this file should now be displayed and the name 'Liver-1' should appear on the samples list.

Step 2 Similarly open the file 'Liver-2'.

Step 3 Select *Combine samples* from the *Special* menu or click on the COMBINE button on the toolbar. Multi-select 'Liver-1' and 'Liver-2' from the samples list (Control-click on PCs or Command-click on Macs).

Step 4 The viewing area should now show the results of combining the 2 data sets. The two experiments give a combined result of 4.638 ± 0.207 μmol/g/h (n = 11).

16.2 Standard error of sums and differences

16.2.1 Uses

We have of course already met the standard error of the difference between two means in considering Student's *t*-test (see Chapter 4). However there are other circumstances where we may wish to know the standard error of the mean. When the mean ± sem for two variables, *p* and *q*, are known the biologically relevant value may be the difference between the two values, or, less frequently, the sum. The standard error of the difference or sum can be calculated from the individual values of sem and n for p and q. Note that the standard error for the sum is the same as the standard error for the difference.

16.2.2 Formulae

Given measurements of two variables p and q with means $\pm$ sem of $\bar{p} \pm sem_p$ $(n = n_p)$ and $\bar{q} \pm sem_q$ $(n = n_q)$, respectively:

Difference between means $= \bar{p} - \bar{q}$

Sum of means $= \bar{p} + \bar{q}$

Standard error of difference between means = standard error of sum of means:

$$= \sqrt{\frac{(n_p - 1).sem_p^2 + (n_q - 1).sem_q^2}{n_p + n_q - 2} \bullet \frac{n_p + n_q}{n_p n_q}}$$

When $n_p = n_q = n$ the above equation reduces to:

$$\sqrt{\frac{sem_p^2 + sem_q^2}{n}}$$

When $n_p \neq n_q$ but both are large – i.e. $n - 1 \approx n$ – the equation reduces to:

$$\sqrt{\frac{sem_p^2}{n_q} + \frac{sem_q^2}{n_p}}$$

Step-by-step tutorial 16.2: standard error of the difference of means

A luciferase-based assay is being used to quantify the amount of ATP and ADP in small tissue samples. The amount of ATP is measured directly in 8 samples as $3.25 \pm 0.14\,\mu\text{mol.g}^{-1}$. A further 10 samples are treated with pyruvate kinase plus phosphoenolpyruvate to convert ADP quantitatively to ATP. The total ATP in these samples is determined to be $4.56 \pm 0.29\,\mu\text{mol.g}^{-1}$. The difference between the two groups gives the mean ADP content as $4.56 - 3.25 = 1.31\,\mu\text{mol.g}^{-1}$.

Step 1 The standard error of the difference in means and hence of the mean ADP content is calculated as:

$$\sqrt{\frac{7 \times 0.14^2 + 9 \times 0.29^2}{8 + 10 - 2} \bullet \frac{8 + 10}{8 \times 10}} = 0.176$$

Step 2 Hence the ADP content is $1.31 \pm 0.176\,\mu\text{mol.g}^{-1}$.

16.3 Standard error of ratios and products

16.3.1 Uses

When the mean ± sem for two variables, p and q, are measured, the physiologically relevant parameter may sometimes be the ratio p/q or, less frequently the product, $p*q$. The *standard errors* of the ratio or product can be calculated from the values of mean, sem and n for p and q.

16.3.2 Formulae

Given measurements of two variables p and q with means ± sem of $\bar{p} \pm sem_p$ ($n = n_p$) and $\bar{q} \pm sem_q$ ($n = n_q$), respectively, the sems of the ratio p/q and the product $p*q$ are given by the following formulae:

Standard error of the ratio

$$SE_{p/q} = \frac{1}{\bar{q}}\sqrt{\frac{n_p.sem_p^2 + n_q\left(\dfrac{\bar{p}}{\bar{q}}\right)^2.sem_q^2}{n_p + n_q - 2}}$$

Standard error of the product

$$SE_{p\times q} = \sqrt{\frac{n_q.\bar{p}^2.sem_q^2 + n_p.\bar{q}^2.sem_p^2 + n_p.sem_p^2.n_q.sem_q^2}{n_p + n_q - 2}}.$$

Step-by-step tutorial 16.3: mean and standard error of ratio

The total protein content was measured in 7 batches of a cell culture. The mean ± sem was 891 ± 22 ng per 10^6 cells. The total DNA content measured in a further 8 samples was 26.0 ± 1.41 ng per 10^6 cells.

Step 1   We wish to know the mean and standard error for the ratio of protein to DNA. This can be calculated as follows:

The mean ratio = 891/26 = 34.27

$$\text{The standard error of the ratio} = \frac{1}{26}\sqrt{\frac{7 \times 1.41^2 + 8 \times \left(\dfrac{891}{26}\right)^2 \times 1.41^2}{8+7-2}}$$

$$= 1.585$$

Step 2   Thus the mean protein:DNA ratio is 34.27 ± 1.585

PractiStat tutorial 16.2: sems of sums, differences, ratios and products

Step 1 A biochemist measured the concentrations of ATP and ADP in a tissue sample as 3.25 ± 0.14 ($n = 8$) and 1.31 ± 0.176 ($n = 10$) μmol.g^{-1} and wishes to know the mean ± sem for the ratio of [ATP]/[ADP].

Step 2 Click on the COMBISTAT button or select *Sums, Products, Ratios of sems* from the *Special* menu to open the CombiStat window.

Step 3 Enter values of mean (3.25), sem (0.14) and n (8) for ATP concentration into the appropriate fields for S1 and values of mean (1.31), sem (0.176) and n (10) for ADP concentration into the appropriate fields for S2 using the TAB key to move from field to field. When all 6 fields have been filled in a further press of the TAB key will generate the values of mean and sem for S1 + S2, S1 − S2, S1/S2 and S1*S2. The ratio [ATP]/[ADP] is shown to be 2.481 ± 0.274.

Step 4 The calculated values can be printed by clicking the Print button. Click on the Exit button to close the CombiStat window and return to the main PractiStat window.

Appendix 1   Critical values for Student's *t*-distribution

df is the degrees of freedom

df		Significance				
	1-tailed:	**0.05**	**0.025**	**0.010**	**0.005**	**0.0005**
	2-tailed:	**0.10**	**0.050**	**0.020**	**0.010**	**0.0010**
1		6.314	12.706	31.821	63.657	636.619
2		2.920	4.303	6.965	9.925	31.599
3		2.353	3.182	4.541	5.841	12.924
4		2.132	2.776	3.747	4.604	8.610
5		2.015	2.571	3.365	4.032	6.869
6		1.943	2.447	3.143	3.707	5.959
7		1.895	2.365	2.998	3.499	5.408
8		1.860	2.306	2.896	3.355	5.041
9		1.833	2.262	2.821	3.250	4.781
10		1.812	2.228	2.764	3.169	4.587
11		1.796	2.201	2.718	3.106	4.437
12		1.782	2.179	2.681	3.055	4.318
13		1.771	2.160	2.650	3.012	4.221
14		1.761	2.145	2.624	2.977	4.140
15		1.753	2.131	2.602	2.947	4.073
16		1.746	2.120	2.583	2.921	4.015
17		1.740	2.110	2.567	2.898	3.965
18		1.734	2.101	2.552	2.878	3.922
19		1.729	2.093	2.539	2.861	3.883
20		1.725	2.086	2.528	2.845	3.850
21		1.721	2.080	2.518	2.831	3.819
22		1.717	2.074	2.508	2.819	3.792
23		1.714	2.069	2.500	2.807	3.767
24		1.711	2.064	2.492	2.797	3.745
25		1.798	2.060	2.485	2.787	3.725
26		1.706	2.056	2.479	2.779	3.707
27		1.703	2.052	2.473	2.771	3.690
28		1.701	2.048	2.467	2.763	3.674
29		1.699	2.045	2.462	2.756	3.659
30		1.697	2.042	2.457	2.750	3.646
40		1.684	2.021	2.423	2.704	3.551
60		1.671	2.000	2.390	2.660	3.460
80		1.644	1.990	2.374	2.639	3.416
100		1.660	1.984	2.364	2.626	3.390
120		1.658	1.980	2.358	2.617	3.373
1000		1.646	1.962	2.330	2.581	3.300
∞		1.645	1.960	2.326	2.576	3.291

Appendix 2   Critical values for Snedecor's *F*-test (2-tailed, 0.05 level of significance)

df₁ and **df₂** are the degrees of freedom for the greater and lesser variances, respectively

df_2 \ df_1	1	2	3	4	5	6	7	8	9	10	12	15	20	24	30	40	60	120	∞
1	647.80	799.50	864.20	899.60	921.80	937.10	948.20	956.70	963.30	968.60	976.70	984.90	993.10	997.20	1001	1006	1010	1014	1018
2	38.51	39.00	39.17	39.25	39.30	39.33	39.36	39.37	39.39	39.40	39.41	39.43	39.45	39.46	39.46	39.47	39.48	39.49	39.50
3	17.44	16.04	15.44	15.10	14.88	14.73	14.62	14.54	14.47	14.42	14.34	14.25	14.17	14.12	14.08	14.04	13.99	13.95	13.90
4	12.22	10.65	9.98	9.60	9.36	9.20	9.07	8.98	8.90	8.84	8.75	8.66	8.56	8.51	8.46	8.41	8.36	8.31	8.26
5	10.01	8.43	7.76	7.39	7.15	6.98	6.85	6.76	6.68	6.62	6.52	6.43	6.33	6.28	6.23	6.18	6.12	6.07	6.02
6	8.81	7.26	6.60	6.23	5.99	5.82	5.70	5.60	5.52	5.46	5.37	5.27	5.17	5.12	5.07	5.01	4.96	4.90	4.85
7	8.07	6.54	5.89	5.52	5.29	5.12	4.99	4.90	4.82	4.76	4.67	4.57	4.47	4.42	4.36	4.31	4.25	4.20	4.14
8	7.57	6.06	5.42	5.05	4.82	4.65	4.53	4.43	4.36	4.30	4.20	4.10	4.00	3.95	3.89	3.84	3.78	3.73	3.67
9	7.21	5.71	5.08	4.72	4.48	4.32	4.20	4.10	4.03	3.96	3.87	3.77	3.67	3.61	3.56	3.51	3.45	3.39	3.33
10	6.94	5.46	4.83	4.47	4.24	4.07	3.95	3.85	3.78	3.72	3.62	3.52	3.42	3.37	3.31	3.26	3.20	3.14	3.08
11	6.72	5.26	4.63	4.28	4.04	3.88	3.76	3.66	3.59	3.53	3.43	3.33	3.23	3.17	3.12	3.06	3.00	2.94	2.88
12	6.55	5.10	4.47	4.12	3.89	3.73	3.61	3.51	3.44	3.37	3.28	3.18	3.07	3.02	2.96	2.91	2.85	2.79	2.72
13	6.41	4.97	4.35	4.00	3.77	3.60	3.48	3.39	3.31	3.25	3.15	3.05	2.95	2.89	2.84	2.78	2.72	2.66	2.60
14	6.30	4.86	4.24	3.89	3.66	3.50	3.38	3.29	3.21	3.15	3.05	2.95	2.84	2.79	2.73	2.67	2.61	2.55	2.49
15	6.20	4.77	4.15	3.80	3.58	3.41	3.29	3.20	3.12	3.06	2.96	2.86	2.76	2.70	2.64	2.59	2.52	2.46	2.40
16	6.12	4.69	4.08	3.73	3.50	3.34	3.22	3.12	3.05	2.99	2.89	2.79	2.68	2.63	2.57	2.51	2.45	2.38	2.32
17	6.04	4.62	4.01	3.66	3.44	3.28	3.16	3.06	2.98	2.92	2.82	2.72	2.62	2.56	2.50	2.44	2.38	2.32	2.25
18	5.98	4.56	3.95	3.61	3.38	3.22	3.10	3.01	2.93	2.87	2.77	2.67	2.56	2.50	2.44	2.38	2.32	2.26	2.19
19	5.92	4.51	3.90	3.56	3.33	3.17	3.05	2.96	2.88	2.82	2.72	2.62	2.51	2.45	2.39	2.33	2.27	2.20	2.13
20	5.87	4.46	3.86	3.51	3.29	3.13	3.01	2.91	2.84	2.77	2.68	2.57	2.46	2.41	2.35	2.29	2.22	2.16	2.09
21	5.83	4.42	3.82	3.48	3.25	3.09	2.97	2.87	2.80	2.73	2.64	2.53	2.42	2.37	2.31	2.25	2.18	2.11	2.04
22	5.79	4.38	3.78	3.44	3.22	3.05	2.93	2.84	2.76	2.70	2.60	2.50	2.39	2.33	2.27	2.21	2.14	2.08	2.00
23	5.75	4.35	3.75	3.41	3.18	3.02	2.90	2.81	2.73	2.67	2.57	2.47	2.36	2.30	2.24	2.18	2.11	2.04	1.97
24	5.72	4.32	3.72	3.38	3.15	2.99	2.87	2.78	2.70	2.64	2.54	2.44	2.33	2.27	2.21	2.15	2.08	2.01	1.94
25	5.69	4.29	3.69	3.35	3.13	2.97	2.85	2.75	2.68	2.61	2.51	2.41	2.30	2.24	2.18	2.12	2.05	1.98	1.91
26	5.66	4.27	3.67	3.33	3.10	2.94	2.82	2.73	2.65	2.59	2.49	2.39	2.28	2.22	2.16	2.09	2.03	1.95	1.88
27	5.63	4.24	3.65	3.31	3.08	2.92	2.80	2.71	2.63	2.57	2.47	2.36	2.25	2.19	2.13	2.07	2.00	1.93	1.85
28	5.61	4.22	3.63	3.29	3.06	2.90	2.78	2.69	2.61	2.55	2.45	2.34	2.23	2.17	2.11	2.05	1.98	1.91	1.83
29	5.59	4.20	3.61	3.27	3.04	2.88	2.76	2.67	2.59	2.53	2.43	2.32	2.21	2.15	2.09	2.03	1.96	1.89	1.81
30	5.57	4.18	3.59	3.25	3.03	2.87	2.75	2.65	2.57	2.51	2.41	2.31	2.20	2.14	2.07	2.01	1.94	1.87	1.79
40	5.42	4.05	3.46	3.13	2.90	2.74	2.62	2.53	2.45	2.39	2.29	2.18	2.07	2.01	1.94	1.88	1.80	1.72	1.64
60	5.29	3.93	3.34	3.01	2.79	2.63	2.51	2.41	2.33	2.27	2.17	2.06	1.94	1.88	1.82	1.74	1.67	1.58	1.48
120	5.15	3.80	3.23	2.89	2.67	2.52	2.39	2.30	2.22	2.16	2.05	1.94	1.82	1.76	1.69	1.61	1.53	1.43	1.31
∞	5.02	3.69	3.12	2.79	2.57	2.41	2.29	2.19	2.11	2.05	1.94	1.83	1.71	1.64	1.57	1.48	1.39	1.27	1.00

Appendix 3   Critical values for Mann–Whitney U-test (2-tailed, 0.05 level of significance)

n_1 and n_2 are the numbers of observations in the larger and smaller samples, respectively

n_2 \ n_1	4	5	6	7	8	9	10	11	12	13	14	15	16	17	18	19	20
2					16	18	20	22	23	25	27	29	31	32	34	36	38
3		15	17	20	22	25	27	30	32	35	37	40	43	45	47	50	52
4	16	19	22	25	28	32	35	38	41	44	47	50	53	57	60	63	67
5		23	27	30	34	37	42	46	49	53	57	61	65	68	72	76	80
6			31	36	39	44	49	53	58	62	67	71	75	80	84	89	93
7				41	46	51	56	61	66	71	76	81	86	91	96	101	106
8					51	57	63	69	74	80	86	91	97	102	108	114	119
9						64	70	76	82	89	95	101	107	114	120	126	132
10							77	84	91	97	104	111	118	125	132	138	145
11								91	99	106	114	121	129	136	143	151	158
12									107	115	123	131	139	147	155	163	171
13										124	132	141	149	158	167	175	184
14											141	151	160	169	178	188	197
15												161	170	180	190	200	210
16													181	191	202	212	222
17														202	213	224	235
18															225	236	248
19																248	261
20																	273

Appendix 4 Critical values for *F*-distribution

df$_1$ and **df$_2$** are the degrees of freedom in the larger and smaller samples, respectively

df$_1$	1		2		3		4		5		6		7	
df$_2$ \ P	0.05	0.01	0.05	0.01	0.05	0.01	0.05	0.01	0.05	0.01	0.05	0.01	0.05	0.01
1	161	4052	199	5000	216	5403	225	5625	230	5764	234	5859	237	5928
2	18.5	98.5	19.0	99.0	19.2	99.2	19.2	99.2	19.3	99.3	19.3	99.3	19.4	99.4
3	10.1	34.1	9.55	30.8	9.28	29.5	9.12	28.7	9.01	28.2	8.94	27.9	8.89	27.7
4	7.71	21.2	6.94	18.0	6.59	16.7	6.39	16.0	6.26	15.5	6.16	15.2	6.09	15.0
5	6.61	16.3	5.79	13.3	5.41	12.1	5.19	11.4	5.05	11.0	4.95	10.7	4.88	10.5
6	5.99	13.7	5.14	10.9	4.76	9.78	4.53	9.15	4.32	8.75	4.18	8.47	4.31	8.26
7	5.59	12.2	4.74	9.55	4.35	8.45	4.12	7.85	3.97	7.46	3.87	7.19	3.79	6.99
8	5.32	11.3	4.46	8.65	4.07	7.59	3.84	7.01	3.69	6.63	3.58	6.37	3.50	6.18
9	5.12	10.6	4.26	8.02	3.86	6.99	3.63	6.42	3.48	6.06	3.37	5.80	3.29	5.61
10	4.96	10.0	4.10	7.56	3.71	6.55	3.48	5.99	3.33	5.64	3.22	5.39	3.14	5.20
11	4.84	9.65	3.98	7.21	3.59	6.22	3.36	5.67	3.20	5.32	3.09	5.07	3.01	4.89
12	4.75	9.33	3.89	6.93	3.49	5.95	3.26	5.41	3.11	5.06	3.00	4.82	2.91	4.64
13	4.67	9.08	3.81	6.70	3.41	5.74	3.18	5.21	3.03	4.86	2.92	4.62	2.83	4.44
14	4.60	8.86	3.74	6.51	3.34	5.56	3.11	5.04	2.96	4.70	2.90	4.46	2.76	4.28
15	4.54	8.68	3.68	6.36	3.29	5.42	3.06	4.89	2.90	4.56	2.79	4.32	2.71	4.14
16	4.49	8.53	3.63	6.23	3.24	5.29	3.01	4.77	2.85	4.44	2.74	4.20	2.66	4.05
17	4.45	8.40	3.59	6.11	3.20	5.18	2.96	4.67	2.81	4.34	2.70	4.10	2.62	3.93
18	4.41	8.29	3.55	6.01	3.16	5.09	2.93	4.58	2.78	4.25	2.66	4.01	2.58	3.84
19	4.38	8.18	3.52	5.93	3.13	5.01	3.90	4.50	2.74	4.17	2.63	3.94	2.54	3.77
20	4.35	8.10	3.49	5.85	3.10	4.94	2.87	4.43	2.71	4.10	2.60	3.87	2.51	3.70
21	4.32	8.02	3.47	5.78	3.07	4.87	2.84	4.37	2.68	4.04	2.57	3.81	2.49	3.64
22	4.30	7.95	3.44	5.72	3.05	4.82	2.82	4.31	2.66	3.99	2.55	3.76	2.46	3.59
23	4.38	7.88	3.42	5.66	3.03	4.76	2.80	4.26	2.64	3.94	2.53	3.71	2.44	3.54
24	4.26	7.82	3.40	5.61	3.01	4.72	2.78	4.22	2.62	3.90	2.51	3.67	2.42	3.50
25	4.24	7.77	3.39	5.57	2.99	4.68	1.76	4.18	2.60	3.89	2.49	3.63	2.40	3.46
26	4.13	7.72	3.37	5.53	2.98	4.64	2.76	4.14	2.59	3.82	2.47	3.59	2.39	3.42
28	4.20	7.64	3.35	5.45	2.91	4.57	2.71	4.07	2.56	3.75	2.45	3.53	2.36	3.36
30	4.17	7.56	3.32	5.39	2.92	4.51	2.69	4.02	2.53	3.70	2.42	3.47	2.33	3.30
40	4.08	7.31	3.23	5.18	2.84	4.31	2.61	3.83	2.45	3.51	2.34	3.29	2.45	3.12
60	4.00	7.08	3.15	4.98	2.76	4.13	2.53	3.65	2.37	3.34	2.25	3.12	2.17	2.95
120	3.92	6.85	3.07	4.79	2.68	3.95	2.45	3.48	2.19	3.17	2.18	2.96	2.09	2.79
∞	3.84	6.63	3.00	4.61	2.60	3.78	2.27	3.32	2.21	3.02	2.10	2.80	2.01	2.64

8		10		12		15		20		40		∞	
0.05	0.01	0.05	0.01	0.05	0.01	0.05	0.01	0.05	0.01	0.05	0.01	0.05	0.01
239	5981	242	6056	244	6106	246	6157	248	6209	251	6287	254	6366
19.4	99.4	19.4	99.4	19.4	99.4	19.4	99.4	19.4	99.4	19.5	99.5	19.5	99.5
8.85	27.5	8.79	27.2	8.74	27.1	8.70	26.9	8.66	26.7	8.59	26.4	8.53	26.1
6.04	14.8	5.96	14.5	5.91	14.4	5.86	14.2	5.80	14.0	5.72	13.7	5.63	13.5
4.82	10.3	4.74	10.1	4.68	9.89	4.62	9.72	4.56	9.55	4.46	9.29	4.36	9.02
4.15	8.10	4.06	7.87	4.00	7.72	3.94	7.56	3.87	7.40	3.78	7.14	3.67	6.88
3.73	6.84	3.64	6.62	3.57	6.47	3.51	6.31	3.44	6.16	3.34	5.91	3.23	5.65
3.44	6.03	3.35	5.81	3.28	5.67	3.22	5.52	3.15	5.36	3.04	5.12	2.93	4.86
3.23	5.47	3.14	5.26	3.07	5.11	3.01	4.96	2.94	4.81	2.83	4.57	2.71	4.31
3.07	5.06	2.98	4.85	2.91	4.71	2.84	4.56	2.77	4.41	2.66	4.17	2.54	3.91
2.95	4.74	2.85	4.54	2.79	4.40	2.72	4.25	2.65	4.10	2.53	3.86	2.40	3.60
2.85	4.50	2.75	4.30	2.69	4.16	2.62	4.01	2.54	3.86	2.43	3.62	2.30	3.36
2.77	4.30	2.67	4.10	2.60	3.96	2.53	3.81	2.46	3.66	2.34	3.43	2.21	3.17
2.70	4.14	2.60	3.94	2.53	3.80	2.46	3.66	2.39	3.51	2.27	3.27	2.13	3.00
2.64	4.00	2.54	3.80	2.48	3.67	2.40	3.52	2.33	3.37	2.20	3.13	2.07	2.87
2.59	3.89	2.49	3.69	2.42	3.55	2.35	3.41	2.28	3.26	2.15	3.02	2.01	2.75
2.55	3.79	2.45	3.59	2.38	3.46	2.31	3.31	2.23	3.16	2.10	2.92	1.96	2.65
2.51	3.71	2.41	3.51	2.34	3.37	2.27	3.23	2.19	3.08	2.06	2.84	1.92	2.57
2.48	3.63	2.38	3.43	2.31	3.30	2.23	3.15	2.16	3.00	2.03	2.76	1.88	2.49
2.45	3.56	2.35	3.37	2.28	3.23	2.20	3.09	2.12	2.94	1.99	2.69	1.84	2.42
2.42	3.51	2.32	3.31	2.25	3.17	2.18	3.03	2.10	2.88	1.96	2.64	1.81	2.36
2.40	3.45	2.30	3.26	2.23	3.12	2.15	2.98	2.07	2.83	1.94	2.58	1.78	2.31
2.37	3.41	2.27	3.21	2.20	3.07	2.13	2.93	2.05	2.78	1.91	2.54	1.76	2.26
2.36	3.36	2.25	3.17	2.18	3.03	2.11	2.89	2.03	2.74	1.89	2.49	1.73	2.21
2.37	3.32	2.24	3.13	2.16	2.99	2.09	2.85	2.01	2.70	1.87	2.45	1.71	2.17
2.32	3.29	2.22	3.09	2.15	2.96	2.07	2.81	1.99	2.66	1.85	2.42	1.69	2.13
2.29	3.23	2.19	3.03	2.12	2.90	2.04	2.75	1.96	2.60	1.82	2.35	1.65	2.06
2.27	3.17	2.16	2.98	2.09	2.84	2.01	2.70	1.93	2.55	1.79	2.30	1.62	2.01
2.18	2.89	2.08	2.80	2.00	2.66	1.92	2.52	1.84	2.37	1.69	2.11	1.51	1.80
2.10	2.72	1.99	2.63	1.92	2.50	1.84	2.35	1.75	2.20	1.59	1.94	1.39	1.60
2.02	2.56	1.91	2.47	1.83	2.34	1.75	2.19	1.66	2.03	1.50	1.76	1.25	1.38
1.94	2.41	1.83	2.32	1.75	2.18	1.67	2.04	1.57	1.88	1.39	1.59	1.00	1.00

Appendix 5 Values of *q* for Tukey test (0.05 level of significance)

m is the number of means being compared and df is the within samples degrees of freedom

df \ m	2	3	4	5	6	7	8	9	10	11	12	13	14	15	16	17	18	19	20
1	17.97	26.98	32.82	37.08	40.41	43.12	45.40	47.36	49.07	50.59	51.96	53.20	54.33	55.36	56.32	57.22	58.04	58.83	59.56
2	6.08	8.33	9.80	10.88	11.74	12.44	13.03	13.54	13.99	14.39	14.75	15.08	15.38	15.65	15.91	16.14	16.37	16.57	16.77
3	4.50	5.91	6.82	7.50	8.04	8.48	8.85	9.18	9.46	9.72	9.95	10.15	10.35	10.52	10.69	10.84	10.98	11.11	11.24
4	3.93	5.04	5.76	6.29	6.71	7.05	7.35	7.60	7.83	8.03	8.21	8.37	8.52	8.66	8.79	8.91	9.03	9.13	9.23
5	3.64	4.60	5.22	5.67	6.03	6.33	6.58	6.80	6.99	7.17	7.32	7.47	7.60	7.72	7.83	7.93	8.03	8.12	8.21
6	3.46	4.34	4.90	5.30	5.63	5.90	6.12	6.32	6.49	6.65	6.79	6.92	7.03	7.14	7.24	7.34	7.43	7.51	7.59
7	3.34	4.16	4.68	5.06	5.36	5.61	5.82	6.00	6.16	6.30	6.43	6.55	6.66	6.76	6.85	6.94	7.02	7.10	7.17
8	3.26	4.04	4.53	4.89	5.17	5.40	5.60	5.77	5.92	6.05	6.18	6.29	6.39	6.48	6.57	6.65	6.73	6.80	6.87
9	3.20	3.95	4.41	4.76	5.02	5.24	5.43	5.59	5.74	5.87	5.98	6.09	6.19	6.28	6.36	6.44	6.51	6.58	6.64
10	3.15	3.88	4.33	4.65	4.91	5.12	5.30	5.46	5.60	5.72	5.83	5.93	6.03	6.11	6.19	6.27	6.34	6.40	6.47
11	3.11	3.82	4.26	4.57	4.82	5.03	5.20	5.35	5.49	5.61	5.71	5.81	5.90	5.98	6.06	6.13	6.20	6.27	6.33
12	3.08	3.77	4.20	4.51	4.75	4.95	5.12	5.27	5.39	5.51	5.61	5.71	5.80	5.88	5.95	6.02	6.09	6.15	6.21
13	3.06	3.73	4.15	4.45	4.69	4.88	5.05	5.19	5.32	5.43	5.53	5.63	5.71	5.79	5.86	5.93	5.99	6.05	6.11
14	3.03	3.70	4.11	4.41	4.64	4.83	4.99	5.13	5.25	5.36	5.46	5.55	5.64	5.71	5.79	5.85	5.91	5.97	6.03
15	3.01	3.67	4.08	4.37	4.59	4.78	4.94	5.08	5.20	5.31	5.40	5.49	5.57	5.65	5.72	5.78	5.85	5.90	5.96
16	3.00	3.65	4.05	4.33	4.56	4.74	4.90	5.03	5.15	5.26	5.35	5.44	5.52	5.59	5.66	5.73	5.79	5.84	5.90
17	2.98	3.63	4.02	4.30	4.52	4.70	4.86	4.99	5.11	5.21	5.31	5.39	5.47	5.54	5.61	5.67	5.73	5.79	5.84
18	2.97	3.61	4.00	4.28	4.49	4.67	4.82	4.96	5.07	5.17	5.27	5.35	5.43	5.50	5.57	5.63	5.69	5.74	5.79
19	2.96	3.59	3.98	4.25	4.47	4.65	4.79	4.92	5.04	5.14	5.23	5.31	5.39	5.46	5.53	5.59	5.65	5.70	5.75
20	2.95	3.58	3.96	4.23	4.45	4.62	4.77	4.90	5.01	5.11	5.20	5.28	5.36	5.43	5.49	5.55	5.61	5.66	5.71
24	2.92	3.53	3.90	4.17	4.37	4.54	4.68	4.81	4.92	5.01	5.10	5.18	5.25	5.32	5.38	5.44	5.49	5.55	5.59
30	2.89	3.49	3.85	4.10	4.30	4.46	4.60	4.72	4.82	4.92	5.00	5.08	5.15	5.21	5.27	5.33	5.38	5.43	5.47
40	2.86	3.44	3.79	4.04	4.23	4.39	4.52	4.63	4.73	4.82	4.90	4.98	5.04	5.11	5.16	5.22	5.27	5.31	5.36
60	2.83	3.40	3.74	3.98	4.16	4.31	4.44	4.55	4.65	4.73	4.81	4.88	4.94	5.00	5.06	5.11	5.15	5.20	5.24
120	2.80	3.36	3.68	3.92	4.10	4.24	4.36	4.47	4.56	4.64	4.71	4.78	4.84	4.90	4.95	5.00	5.04	5.09	5.13
∞	2.77	3.31	3.63	3.86	4.03	4.17	4.29	4.39	4.47	4.55	4.62	4.68	4.74	4.80	4.85	4.89s	4.93s	4.97	5.01

Appendix 6 Critical values for Chi-square-distribution

df is the degrees of freedom and α is the significance level

df \ α	0.05	0.01	0.001
1	3.84	6.64	10.83
2	5.99	9.21	13.82
3	7.82	11.35	16.27
4	9.49	13.28	18.47
5	11.07	15.09	20.52
6	12.59	16.81	22.46
7	14.07	18.48	24.32
8	15.51	20.09	26.13
9	16.92	21.67	27.88
10	18.31	23.21	29.59
11	19.68	24.73	31.26
12	21.03	26.22	32.91
13	22.36	27.69	34.53
14	23.69	29.14	36.12
15	25.00	30.58	37.70
16	26.30	32.00	39.25
17	27.59	33.41	40.79
18	28.87	34.81	42.31
19	30.14	36.19	43.82
20	31.41	37.57	45.32
21	32.67	38.93	46.80
22	33.92	40.29	48.27
23	35.17	41.64	49.73
24	36.42	42.98	51.18
25	37.65	44.31	52.62
26	38.89	45.64	54.05
27	40.11	46.96	55.48
28	41.34	48.28	56.89
29	42.56	49.59	58.30
30	43.77	50.89	59.70
40	55.76	63.69	73.40
50	67.51	76.15	86.66
60	79.08	88.38	99.61
70	90.53	100.42	112.32
80	101.88	112.33	124.84
90	113.15	124.12	137.21
100	124.34	135.81	149.45

Appendix 7 Critical values for Spearman's Rank Correlation Coefficient

n is the number of observations in each of the two series

	Significance			
n	**1-tailed:** 0.05 **2-tailed:** 0.10	0.025 0.050	0.010 0.020	0.005 0.010
4	1.000	—	—	—
5	0.900	—	—	—
6	0.829	0.886	0.943	—
7	0.714	0.786	0.893	—
8	0.643	0.738	0.833	0.881
9	0.600	0.683	0.783	0.833
10	0.564	0.648	0.745	0.794
11	0.523	0.623	0.736	0.818
12	0.497	0.591	0.703	0.780
13	0.475	0.566	0.673	0.745
14	0.457	0.545	0.646	0.716
15	0.441	0.525	0.623	0.689
16	0.425	0.507	0.601	0.666
17	0.412	0.490	0.582	0.645
18	0.399	0.476	0.564	0.625
19	0.388	0.462	0.549	0.608
20	0.377	0.450	0.534	0.591
21	0.368	0.438	0.521	0.576
22	0.359	0.428	0.508	0.562
23	0.351	0.418	0.496	0.549
24	0.343	0.409	0.485	0.537
25	0.336	0.400	0.475	0.526
26	0.329	0.392	0.465	0.515
27	0.323	0.385	0.456	0.505
28	0.317	0.377	0.448	0.496
29	0.311	0.370	0.440	0.487
30	0.305	0.364	0.432	0.478

Glossary

1-tailed test A test that rejects outcomes in only one specified tail of a given distribution

1-way ANOVA *ANOVA* in which the groups are defined by a single *variable*. Described in Chapter 8

2-tailed test A test that rejects extreme values at either side of the distribution

2-way ANOVA *ANOVA* for comparing the *means* of several groups for the effects of two *independent variables*; described in Chapter 13

A priori tests *Multiple comparison tests* which are made on only a subset of the data, as decided before the experiment is performed

Abscissa The horizontal (x)-axis of a 2-dimensional graph

Alpha (α) Symbol used to denote the probability of making a *Type I error*

Analysis of variance A method for testing for significant differences between *means* by comparing *variances*

ANOVA A contraction of *analysis of variance*

Approximate t-test A modified form of *Student's t-test* used when the *variances* of the two *samples* are significantly different; described in Chapter 4

Best straight line The straight line drawn through a set of data of data points $[x_i, y_i]$ which minimises the the sum of the squared deviations of the experimental values from the fitted line (the *residual sum of squares*)

Beta (β) Symbol used to denote the probability of making a *Type II error*

Bimodal Describes a distribution having two distinct peaks

Bonferroni inequality A correction applied to a probability level to take account of the increased likelihood of making a *Type I error* when carrying out multiple comparisons

Bonferroni's t-test A *post hoc test* to determine which means show significant differences after *analysis of variance* has indicated that a statistically significant difference exists; described in Chapter 10

Box plot A graphical representation of data in which *ranges* and distribution characteristics of values of a selected *variable* are plotted; described in Chapter 3

Categorical variable See *Qualitative variable*

Chi-square distribution The distribution of the χ^2-statistic

Chi square test A statistical test applicable to the analysis of *qualitative data*; described in Chapter 15

Class interval A set interval chosen for assigning data when plotting a *histogram*

Coefficient of variation The *standard deviation* expressed as a percentage of the *mean*

Combinations Ways in which a given set of objects can be arranged when order is not taken into account

Confidence interval A range of values around a *statistic* where the true (*population*) *statistic* can be expected to be situated with a specified level of certainty

Contingency coefficient A coefficient indicating the degree of relationship exhibited by a *contingency table*

Contingency table A table in which each observation is classified simultaneously in terms of two *variables*

Continuous variable A *variable* that can take any numerical value

Control *t*-test A *post hoc method* for comparing the *mean* of one control sample against that of several other samples after an *analysis of variance* has indicated that a statistically significant difference exists; described in Chapter 10

Correlation coefficient A measure of the strength of the relationship between two *variables*

Correlation A method for determining whether two *variables* have a linear relationship and for quantifying the strength of the relationship

Critical value The value of a *test statistic* at or beyond which the *null hypothesis* will be rejected

Curvilinear relationship A relationship between two *variables* that cannot be represented as a straight line

Degrees of freedom The number of values in a set of observations which can be assigned arbitrarily within the specification of the system; a group of n observations being used to estimate a single *population* characteristic has n-1 degrees of freedom

Dependent variable A *variable* that is measured in contrast to an *independent variable* which is controlled by the experimenter

Descriptive statistics *Statistics* which describe the characteristics of the sample data; described in Chapter 2

Discrete variables *Variables* that can take only a limited number of values

Distribution free test A test that does not rely on estimation of *parameters* or assumptions about distribution; synonymous with *Non-parametric test*

Dot plot A graphical representation of data in which the frequencies of individual values are plotted as points stacked above the *x*-axis; described in Chapter 3

Expected frequencies Number of occurrences expected if a particular hypothesis is true

Expected value The average value of a *statistic* when calculated for an infinite number of *samples*

F-distribution The distribution of the ratio calculated in a *Variance ratio test*

Fisher Least Significant Difference (LSD) test A *post hoc method* to determine which *means* show the significant differences after *analysis of variance* has indicated that a statistically significant difference exists; described in Chapter 10

Frequency distribution A distribution in which the values of the *dependent variable* are plotted against their frequency of occurrence

Friedman test A *non-parametric test* for comparing several *matched samples*; described in Chapter 12

Goodness of fit test A test for comparing observed frequencies with those predicted by a particular hypothesis; described in Chapter 15

Histogram A graphical representation of the *frequency distribution* of a *variable* in which the columns are drawn over the class intervals and the heights of the columns are proportional to the class frequencies; described in Chapter 3

Independent events Events whose outcomes do not influence each other

Independent variable A variable that is controlled by the experimenter in contrast to a *dependent variable* which is only measured

Interaction Modification of a relationship between (at least) two *variables* by (at least) one other *variable*.

Intercept On a two-dimensional (*X-Y*) graphic plot, the value of *Y* when $X = 0$

Interpolation Projecting a curve or distribution between known values to infer the value of a *variable* at intermediate points

Interquartile range (IQR) The *range* of the middle 50% of a set of ranked data

Kruskal-Wallis test A non-parametric test used to compare the *means* of three or more *samples*; described in Chapter 11

Kurtosis A measure of the peakedness of a distribution

Least squares line See *regression line*

Linear regression *Regression* in which the relationship is linear; described in Chapter 14

Linear relationship A relationship for which the best fit is to a straight line

Mann-Whitney U-test A *non-parametric test* for comparing the *medians* of two *samples*; described in Chapter 5

Matched pairs An experimental design in which each subject is tested twice e.g. before and after a treatment

Matched samples ANOVA Analysis of variance where the data are *matched pairs* and the number of conditions is three or more; described in Chapter 9

Matched samples t-test A special version of Student's *t*-test used when the data consist of *matched pairs*; either a certain number of subjects are each measured under two conditions, A and B, or measurements are carried out on one member of a pair under condition A and on the other member of the pair under condition *B*; described in Chapter 6

Mean   The arithmetic average of a number of observations

Measure of central tendency   Numerical values referring to the centre of a distribution – *mean, median and mode*; described in Chapter 2

Median   The middle value of a set of *ranked data*

Mode   The most commonly occurring value in a set of observations

Multiple comparison tests   *Post hoc methods* for comparing *means* of several groups; described in Chapter 10

Multiple regression   *Regression* with two or more *variables*

Negatively skewed   A distribution that has a relatively large number of data occurring to the left of (more negative than) the *mean*

Nominal variables   Variables which allow for only qualitative classification. Typical examples are gender, race, blood group

Non-directional test   See *2-tailed test*

Non-parametric tests   Tests for statistical significance which do not rely on assumptions concerning the distribution of the *variable* of interest in the *population*

Normal curve   See *normal distribution*

Normal distribution   A bell-shaped symmetrical curve found to approximate to the *frequency distribution* often obtained when large samples of measurements of a *parameter* are made

Normalised value   See *z-score*

Normality   The assumption that the data being sampled follows the *normal distribution*

Normality test   A test to verify the assumption of *normality* before using a *parametric test*

Null hypothesis (H_0)   The hypothesis tested by a particular statistical procedure, usually expressed in terms of there being no difference between two *populations*

Observed frequencies   The number of occurrences actually observed in contrast to those predicted by a hypothesis

Odds   The ratio of the probability of an event occurring to that of it not occurring

Ordinal variables   A *variable* which can be ranked in terms of which has less and which has more of the quality represented by the variable, but does not allow us to quantify 'how much more' e.g. the pain of an injection

Ordinate   The vertical (Y) axis

Outlier threshold   A value above or below which an observation is regarded as an *outlier*

Outlier   An extreme data point which does not appear to follow the characteristic distribution of the rest of the data

P-value   The probability of error involved in accepting an observed result as valid i.e. as representative of the *population*. For example, a P-level of

0.05 indicates that there is a 5% probability that the apparent relation between the *variables* found in our *sample* is due to chance

Paired samples See *Matched pairs*

Paired *t*-test See *Matched samples t-test*

Parameter Numerical value summarising some aspect of a set of data

Parametric tests Statistical tests that involve estimation and/or assumptions about *parameters* of the distribution

Pearson's correlation coefficient The most widely used type of *correlation coefficient*; described in Chapter 14

Percentile A number below which a specified percentage of the population values fall e.g. the 25th percentile of a variable is a value such that 25% of the values of the variable fall below that value

Population The complete set of all possible values of a *variable*

Population mean The *mean* of the *population*, usually estimated from observations on a *sample*

Population standard deviation The *standard deviation* of the *population*, usually estimated from observations on a *sample*

Population variance The *variance* of the *population*, usually estimated from observations on a *sample*

Positively skewed A distribution that has a relatively large number of data occurring to the right of (more positive than) the *mean*

Post hoc tests Methods used after obtaining a statistically significant result from *ANOVA* to determine which *means* contributed to the effect i.e. which groups are significantly different from each other; described in Chapter 10

Power The probability of correctly rejecting a false *null hypothesis*

Power calculations Methods for determining the minimum number of observations required in an experiment in order to achieve a desired *power* and *significance level*

Probability A measure of the likelihood of something occurring

Product moment correlation coefficient See *Pearson correlation coefficient*

Qualitative variable Sometimes referred to as a *categorical variable*, their values can be non-numerical and assume the value of the category to which they belong. *Ordinal variables* and *Nominal variables* are the two types of qualitative variable

Quantiles Generic name of parameters such as *percentiles* and *quartiles*

Quantitative variable A variable for which the data have been obtained by measurement which can be assigned a numerical value of a standard measurement unit (e.g., inches, grams, litres, etc.)

Quartile Values signifying the upper bounds of the first, second and third quarters of a set of ranked data

Random sample A group of observations obtained from a *population* in a manner that gave each member of the *population* an equal chance of being selected

Randomised blocks design An experimental design for *Matched-samples ANOVA* in which the subjects in each matched set are randomly assigned to one or another of the conditions of measurement

Range   The difference between the highest and the lowest value of a set of observations

Rank order   The position of an observation in a list of values that has been sorted in order with the lowest value being assigned a rank of 1

Ranked data   Data whose numerical values have been replaced by their *rank order*

Regression   A method for estimating the value of one *variable* from knowledge of one or more other variables

Regression analysis   A method for determining the relationship between one or more *independent variables* and a single *dependent variable*

Regression coefficient   The slope of a *regression line*

Regression equation   An equation that allows you to predict a value for a dependent variable given a value or values for one or more independent variables

Regression line   The line that best fits plotted data

Repeated measures design   An experimental design for *Matched-samples ANOVA* in which each subject is measured under each of the experimental conditions. Also called *Within-subjects design*

Residual sum of squares   The sum of the squares of the *residuals*

Residuals   Differences between observed values and their corresponding values predicted by a *regression*

Robust   Descriptive of a test that is not seriously compromised if its underlying assumptions are not fully met

Sample   The actual observations made and comprising a subset of the *population*

Sample mean   The *mean* value of a *sample*

Sample standard deviation   The square root of the *sample variance*

Sample statistics   Statistics calculated from *sample parameters*

Sample variance   Sum of the *squared deviations* from the *mean* of the sample divided by n-1

Sampling distribution   The distribution of a *statistic* when repeated measures of that statistic are made

Sampling error   Chance variability of a *statistic* from *sample* to sample

Scatter plot   A 2-dimensional plot of individual data points

Sigma (σ)   Symbol for *standard deviation* of a *population*

Sigma (Σ)   Symbol indicating summation

Significance level   The probability of incorrectly rejecting a true *null hypothesis*

Skewness   A measure of the deviation of a distribution from *symmetry* about its *mean*. *Normal distributions* are perfectly symmetrical about their mean i.e. have a skewness of zero

Snedecor's F-test   See *Variance ratio test*

Spearman's rank correlation coefficient   A *correlation coefficient* computed from *ranks*; described in Chapter 14

Standard deviation A commonly used measure of variation equal to the square root of the *variance*

Standard error of the mean The *standard deviation* of all *sample means* of size *n* drawn from a *population* – its value depends on both the *population variance* and *n*

Statistic A numerical value summarising *sample* data

Statistical distribution The range of values of a variable expressed in terms of frequency

Statistical inference Drawing conclusions about *populations* from observations on *samples*

Statistical significance The probability of error involved in accepting an observed result as valid i.e. as representative of the *population*; the probability of rejecting a *null hypothesis* that is in fact correct.

Statistical test A procedure for determining *statistical significance*

Student's *t*-distribution The distribution of Student's *t*-statistic

Student's *t*-test A *parametric* test for comparing the *means* of two independent samples; described in Chapter 4

Sum of squares The sum of the squared deviations from some defined value, usually the *mean*

Symmetric Having identical shape on both sides of the centre

Test statistic The result(s) obtained from a *statistical test*

Tukey test A *post hoc test* for *paired samples* to determine which means show significant differences after *analysis of variance* has indicated that a statistical difference exists; described in Chapter 10

Type I error The error of rejecting a true statistical *null hypothesis*

Type II error The error of accepting a false statistical *null hypothesis*

Unbiased estimate A *statistic* whose *expected value* is equal to that of the *parameter* being estimated

Variable A property of an object that can be ascribed a value

Variance A measure of variability equal to the sum of the squared deviations from the *mean* divided by the number of *degrees of freedom* (*n*-1)

Variance ratio test A test to determine whether the variances of two samples are significantly different also known as *Snedecor's F-test*; described in Chapter 4

Wilcoxon Signed-Rank test A *nonparametric test* for *matched samples*; described in Chapter 7

Within-subjects design See *Repeated measures design*

Yates' correction A correction to adjust *chi-square* in small 2×2 tables by reducing the absolute value of differences between expected and observed frequencies by 0.5 before squaring

Z-distribution A general and fundamental distribution which includes the *normal distribution*, the χ^2 *distribution* and *Student's t-distribution* as special cases

Z-score The number of *standard deviations* that a value is away from the *mean*

Index